T0298324

Smart Ships

Smart Ships

Edited by

Yang Xiao
Tieshan Li

CRC Press
Taylor & Francis Group
Boca Raton London New York

CRC Press is an imprint of the
Taylor & Francis Group, an **informa** business

First edition published 2023
by CRC Press
6000 Broken Sound Parkway NW, Suite 300, Boca Raton, FL 33487-2742

and by CRC Press
4 Park Square, Milton Park, Abingdon, Oxon, OX14 4RN

ISBN: 978-0-367-45830-0 (hbk)
ISBN: 978-1-032-07020-9 (pbk)
ISBN: 978-1-003-02592-4 (ebk)

DOI: 10.1201/9781003025924

Typeset in CMR10 font
by KnowledgeWorks Global Ltd.

Contents

vi   *Contents*

Preface

Smart shipping integrates shipping with many fields such as fishing, manufacturing, navigation, communication, computing, control, sensing, etc., to provide better shipping and services. This book presents state-of-the-art approaches and novel technologies for smart ships covering a range of topics in the areas including general shipping concepts and architecture; ship formation control; energy-saving; artificial intelligence; security and privacy of navigation, localization, and path planning; infectious disease and indoor air quality management; smart Autonomous Underwater Vehicle (AUV); smart spectrum usage; etc. This book is an essential source of reference for professionals and researchers in the areas of fishing, manufacturing, navigation, communication, computing, control, and sensing and serves as a text for graduate students in these fields.

This book is made possible by the great efforts of our contributors and publisher. We are indebted to our contributors, who have sacrificed days and nights to collaboratively write these chapters for our readers during the COVID-19 pandemic. We also would like to thank our publisher. Without their encouragement and expertise, this book would not have been a reality. Finally, we are grateful that our families have continuously supported us.

Yang Xiao
Navigation College, Dalian Maritime University, Dalian, China
Department of Computer Science, The University of Alabama, Tuscaloosa,
AL 35487-0290, USA
E-mail: yangxiao@ieee.org

Tieshan Li
School of Automation Engineering, University of Electronic
Science and Technology of China, Chengdu, China
Navigation College, Dalian Maritime University, Dalian, China
E-mail: tieshanli@126.com

About the Editors

Yang Xiao received the B.S. and M.S. degrees in computational mathematics from Jilin University, Changchun, China, in 1989 and 1991, respectively, and the M.S. and Ph.D. degrees in computer science and engineering from Wright State University, Dayton, OH, USA, in 2000 and 2001, respectively. He is currently a Full Professor with the Department of Computer Science, The University of Alabama, Tuscaloosa, AL, USA. He had directed 20 doctoral dissertations and supervised 19 M.S. theses/projects. He has published over 300 Science Citation Index (SCI)-indexed journal papers (including over 60 IEEE/ACM TRANSACTIONS) and 300 Engineering Index (EI)-indexed refereed conference papers and book chapters related to these research areas. His current research interests include cyber–physical systems, the Internet of Things, security, wireless networks, smart grid, and telemedicine. When this book was prepared, Dr. Xiao was visiting Navigation College, Dalian Maritime University during the summer.

Prof. Xiao was a Voting Member of the IEEE 802.11 Working Group from 2001 to 2004, involving the IEEE 802.11 (Wi-Fi) standardization work. He is an IEEE Fellow, an IET Fellow, and an AAIA Fellow. He has served as a Guest Editor over 30 times for different international journals, including the IEEE TRANSACTIONS ON NETWORK SCIENCE AND ENGINEERING in 2021, IEEE TRANSACTIONS ON GREEN COMMUNICATIONS AND NETWORKING in 2021, IEEE Network in 2007, IEEE WIRELESS COMMUNICATIONS in 2006 and 2021, IEEE Communications Standards Magazine in 2021, and Mobile Networks and Applications (MONET) (ACM/Springer) in 2008. He also serves as the Editor-in-Chief of Cyber-Physical Systems journal, International Journal of Sensor

Networks (IJSNet), and International Journal of Security and Networks (IJSN). He has been serving as an Editorial Board Member or an Associate Editor for 20 international journals, including the IEEE TRANSACTIONS ON CYBERNETICS since 2020, IEEE TRANSACTIONS ON SYSTEMS, MAN, AND CYBERNETICS: SYSTEMS from 2014 to 2015, IEEE TRANSACTIONS ON VEHICULAR TECHNOLOGY from 2007 to 2009, and IEEE COMMUNICATIONS SURVEYS AND TUTORIALS from 2007 to 2014.

Tieshan Li earned his B.S. degree in ocean fisheries engineering from Ocean University of China, Qingdao, China, in 1992, and his Ph.D. degree in vehicle operation engineering from Dalian Maritime University (DMU), Dalian, China, in 2005. He has been a Full Professor with the School of Automation Engineering, University of Electronic Science and Technology of China, Chengdu, China, since 2019. He was a Lecturer with DMU from 2005 to 2006, where he was an Associate Professor from 2006 to 2011, and a Full Professor since 2011. From 2007 to 2010, he was a Postdoctoral Scholar with the School of Naval Architecture, Ocean and Civil Engineering, Shanghai Jiao Tong University, Shanghai, China. From 2008 to 2009, and from 2014 to 2015, he visited the City University of Hong Kong as a Senior Research Associate. Since 2013, he has been Visiting Scholar at the University of Macau, Macau, China. His current research interests include intelligent learning and control for nonlinear systems and multi-agent systems and their applications to marine vehicle control.

Contributors

Eirini Barri
University of Patras
Department of Computer Engineering and Informatics, 26504 Rio Patras, Greece

Christos Bouras
University of Patras,
Department of Computer Engineering and Informatics, 26504 Rio Patras, Greece

C. L. Philip Chen
South China University of Technology,
School of Computer Science and Engineering,
Guangzhou, Guangdong, 510006, China

A. C. M. Fong
Western Michigan University
Department of Computer Science,
1903 W Michigan Ave, Kalamazoo, MI 49008, USA

Bernard Fong
Providence University
College of Computing and Informatics,
Taiwan Blvd Sec. 7 No. 200, Taichung 433, Taiwan

Rachit Garg
Dr. Babasaheb Ambedkar Technological University.
Department of Computer Engineering.
Lonere 402103 Dist Raigad (MS), India.

Zijun Gong
University of Newfoundland,
Faculty of Engineering and Applied Science, Memorial St.,
John's, N.L., A1B 3X5, Canada

Apostolos Gkamas
University Ecclesiastical Academy of Vella,
P.O. Box 1144, 45001, Ioannina, Greece

Fan Jiang
Chalmers University of Technology,
Department of Electrical Engineering,
SE-412 96 Goteborg, Sweden

Nikos Karacapilidis
University of Patras,
Industrial Management and Information Systems Lab, MEAD, 26504 Rio
Patras, Greece

Dimitris Karadimas
OptionsNet S.A, 121 Maizonos Str., 26222, Patras, Greece

Georgios Kournetas
University of Patras,
Industrial Management and Information Systems Lab, MEAD, 26504 Rio
Patras, Greece

Arvind W. Kiwelekar
Dr. Babasaheb Ambedkar Technological University,
Department of Computer Engineering,
Lonere 402103 Dist Raigad (MS), India

Cheng Li
Memorial University of Newfoundland,
Faculty of Engineering and Applied Science,
St. John's, N.L., A1B 3X5, Canada

C. K. Li
Alpha Positive Clinic
23/F, New World Tower 1, 18 Queen's Road, Central, Hong Kong

Jiacheng Li
The University of Alabama,
Department of Computer Science,
Tuscaloosa, AL 35487-0290, USA,

Tieshan Li
University of Electronic Science and Technology of China,
School of Automation Engineering,
Chengdu, 610054, China,

Laxman D. Netak
Dr. Babasaheb Ambedkar Technological University,
Department of Computer Engineering,
Lonere 402103 Dist Raigad (MS), India

Yiannis Panaretou
OptionsNet S.A, 121 Maizonos Str., 26222, Patras, Greece

Hasan Bora Usluer
T.R. Galatasaray University,
Ortakoy, Istanbul, Turkey

Ruoyu Su
Nanjing University of Posts & Telecommunications,
School of Internet of Things, 210023, Nanjing, China

Jia Wang
Dalian Maritime University,
Navigation College, Dalian, Liaoning 116026, China

Yang Xiao
The University of Alabama,
Department of Computer Science,
Tuscaloosa, AL 35487-0290, USA

Yue Yang
Dalian Maritime University,
Navigation College, Dalian, Liaoning 116026, China

Acknowledgment

Tieshan Li's work was partially supported by the National Natural Science Foundation of China (under Grant Nos. 51939001, 61976033); the Science and Technology Innovation Funds of Dalian (under Grant No. 2018J11CY022); the Liaoning Revitalization Talents Program (under Grant Nos. XLYC1908018, XLYC1807046); and the Fundamental Research Funds for the Central Universities (under Grant No. 3132019345).

PART 1
General Overview of Smart Ships

Chapter 1

Ship Architecture and Functionalities

Jia Wang[1,2], Yang Xiao[3,*], Tieshan Li[4,1], C. L. Philip Chen[5,1]

[1] *Navigation College, Dalian Maritime University, Dalian, Liaoning 116026, China.*

[2] *Experimental Instrument Center, Dalian Polytechnic University, Dalian, Liaoning 116034, China (e-mail: dmuwangjia@163.com).*

[3] *Department of Computer Science, The University of Alabama,Tuscaloosa, AL 35487-0290, USA (e-mail: yangxiao@ieee.org).*
* Yang Xiao is the corresponding author.

[4] *School of Automation Engineering, University of Electronic Science and Technology of China, Chengdu, 611731, China (e-mail: tieshanli@126.com).*

[5] *School of Computer Science and Engineering, South China University of Technology, Guangzhou, Guangdong, 510006, China (e-mail: philip.chen@ieee.org).*

1.1. Introduction

In this chapter, we introduce some basic ship structure and functionalities [1], [2]. We took some photos of a ship of Dalian Maritime University, called "Yukun". These photos are used to illustrate the basic structures and functions of a typical ship. More comprehensive studies can be found in [1]. Yukun is a ship with breadth = 18.00 m, length = 116.00 m, designed draft = 5.40 m, depth = 11.10 m, endurance = 10000 n.miles, complement = 236 p, main engine = 4440 kW * 173 rpm, service speed = 16.7 knots, and its classification is Ice Class B. It is a training ship. The equipment and structure of the ship fully comply with the China Classification Society regulations and are navigation suitable in the unlimited navigation area. The manufacture and installation of propulsion machinery and auxiliary machinery for significant objectives of the ship comply with the provisions of the China Classification Society code and apply to navigation in unlimited areas. It is a propeller condition monitoring ship with an engine room automation degree and can be unattended when navigation is going on.

The rest of the chapter is organized as follows. In Section 1.2, we present a typical ship structure. In Section 1.3, ship handling is presented. We introduce navigation instruments in Section 1.4. Finally, we conclude this chapter in Section 7.8.

1.2. Ship Structure

Figure 1.1 shows the composition of a merchant ship including a super-structure, main hull, and all kinds of equipment. The main hull is made of hollow watertight construction, which consists of the lower deck, upper deck, wide side, bulkhead, front and rear deck, etc. The superstructure is composed of poop deck, deckhouse, forecastle, and navigating bridge. In addition to various functional cabins in the superstructure, each deck and bulkhead and several cabins separate the main hull into a cargo room, engine room, deep tank, ballast tank, lubricating oil tank, fuel oil tank, slop tank, caisson, and freshwater tank. The supporting equipment of a merchant ship ensures safe navigation, including an auxiliary engine, main engine, electricity, various pipes, auxiliary, safety, deck, communication, living, and navigation equipment.

Fig. 1.1.   A navigation ship "Yukun" model of Dalian Maritime University

Fig. 1.2.   Main engine

The main engine shown in Figure 1.2 and other auxiliary equipment are in the engine room of a ship [4]. The main engine of a ship has four categories: steam engines, internal combustion engines, nuclear power engines, and motors from the aspects of working mode, burning place, and nature of the fuel. In the 20th century, most ships used mechanical propulsion when sailing on water [1]. With the ship propulsion technology's continuous development, some new propulsion technologies include water jet, pump-jet, air cushion, azimuth, and so on. These technologies not only improve the speed of ships, but also improve the efficiency.

The steering gear is often installed at the stern and is the main equipment for controlling the ship's navigation. Its function is to introduce a huge moment of turning to maintain the required course, change the original course, and carry out a cyclic movement, as shown in Figure 1.3. The steering gear is normally installed at the stern peak deck platform of the steering gear cabin. The steering control system transmits the steering command from the crew in the bridge to the steering engine via the hydraulic or electric control system to control the ship's heading and direction.

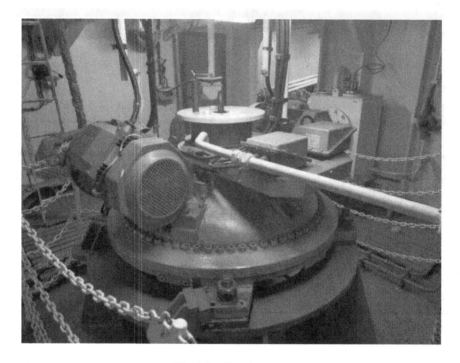

Fig. 1.3. Steering gear

In terms of usage, the piping systems include bilge, sanitary water, ballast pipe, fire extinguishing, deck scupper, domestic water supply, ventilating, etc.

An anchor is the main component of the anchoring equipment used for auxiliary control and ship anchoring of ship berthing, navigation deceleration in a narrow channel, as shown in Figure 1.4. The windlass, as shown in Figure 1.5, is a mechanical device that can be operated remotely from the bridge to drop and heave an anchor. It is composed of a hydraulic and an electric anchor located at the bow of a ship.

1.3. Ship Handling

Ship handling enables a ship to keep or change its horizontal motion, using the hull's relative motion, propeller, and rudder in the water. The crew shall keep or change the speed, position, and course of a ship according to the actual situation.

Fig. 1.4. Anchor

The autopilot can keep the ship heading for a long time and correct the yaw in terms of time, which can reduce the working stress of the helmsman. Large ships on the sea are usually equipped with an autopilot which automatically controls the course. An autopilot consists of automatic detection of signal comparison, course deviation, feedback, actuator, signal amplification, and so on. A gyrocompass detects whether the ship deviates from the route when a ship uses autopilot.

Figure 1.6 shows a compass that is used by many navigation ships. The yaw angle ϕ will be detected by a gyrocompass when the ship goes off course. AC voltage is produced proportionally to a yaw angle by an

Fig. 1.5.   Windlass

automatic steering transmitter to obtain DC voltage U_ϕ of different polarity from phase-sensitive rectifying. A gyrocompass sends a voltage signal to a comparison circuit, which uses an amplifier and a switching circuit to start the motor to manipulate the rudder left or right. Many ships use autopilot to decrease the number of rudder rotation and resistance, maintain the ship on the old course, and reduce fuel consumption.

A navpilot, which is automatic navigation, uses an autopilot and connects Global Navigation Satellite System (GNSS) receivers and integrated navigators to provide reliable course control to the ship. In particular, manual handling should be used in case of bad traffic or bad weather.

Ships need to drop an anchor or anchors when they are prepared for shelter, anchorage operation, and berthing from the wind. Usually, a chief officer is responsible for the handling of an anchor and conducting the bridge's anchoring order. Moreover, the current status and the execution time of the chain of the anchor during anchoring must be reported to the bridge by the chief officer.

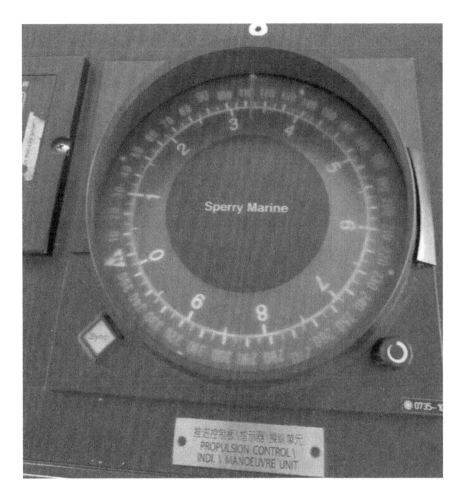

Fig. 1.6. Compass indicator

1.4. Navigation Instruments

The trend of ship development is toward large-scale, high-speed, and intensive navigation to cope with the complicated navigation environment. Navigation instruments must be widely used in ships to ensure the safety of navigation. We explain some widely adopted navigation instruments as follows.

The magnetic needle of the magnetic compass is a kind of navigation pointing instrument, always pointing to the earth's North Pole through the geomagnetic field's mutual attraction. The course and position of ships are

Fig. 1.7.   Marine echo sounder

indicated by a magnetic compass to help ships implement the positioning and navigation functions. The safe navigation of ships is guaranteed by both a magnetic compass and gyrocompass. At present, researchers have developed a solid-state electronic magnetic compass with a magnetic sensor as the core, which converts the induction of external magnetic field from a magnetic sensor into electrical signals. At present, there are several kinds of compasses, such as hall, magnetoresistive, fluxgate, giant magnetoresistive, etc.

Figure 1.7 shows a marine echo sounder for acoustic navigation using water's ultrasonic propagation to measure water's depth. A marine echo

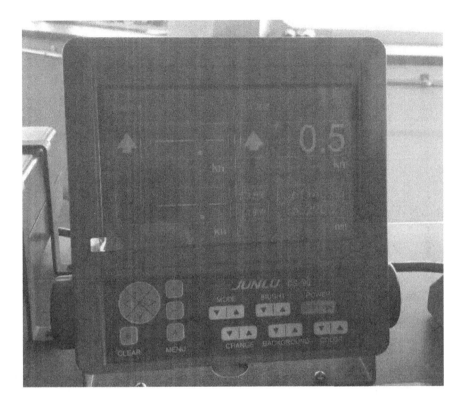

Fig. 1.8. Speed log

sounder determines water's depth via measuring reflected ultrasonic waves and emitted ultrasonic waves, as well as time intervals between them.

A log for speed is an important navigation device, which can measure the accumulated voyage and the speed of a ship, as shown in Figure 1.8. An electromagnetic log measures the ship's voyage and the ship's speed relative to the water-based on the electromagnetic induction principle, including an electromagnetic sensor, indicator, and amplifier. Moreover, a speed log has many advantages in good linearity, large range, low cost, high precision, ease of use, and so on.

The GNSS can provide all-weather 3D coordinates, speed, and time information for ships, including the Global Positioning System (GPS) of the United States, Global Navigation Satellite System (GLONASS) of Russia, Galileo of Europe, BeiDou Navigation Satellite System of China, etc. At present, GPS is widely used in marine navigation, so we will focus on GPS

Fig. 1.9.   DGPS

as follows. GPS satellite navigator is a GPS receiver used for navigation
and positioning. Difference GPS (DGPS) adds correction signals based on
GPS to improve the accuracy, which is used for route navigation, track line
plotting, alarm, fixed-point navigation, as shown in Figure 1.9.

A marine radar (radio detection and ranging) is the X-band or the
S-band radar onboard which detects the target by transmitting electromagnetic waves and receiving the reflected echo of the target. The radar
includes a timer, transmitter, transceiver switch, power supply, receiver,
antenna, display, etc. Radar can detect surface vessels, obstacles, sea clutter, land, and maritime distress to avoid a collision, navigation, and positioning [5], as shown in Figure 1.10. At present, radar/ARPA (Automatic
Radar Plotting Aid) is an important part of the automatic navigation system on ships.

An Electronic Chart Display and Information System (ECDIS) is an
advanced information system for nautical ship navigation [6], as shown in
Figure 1.11. The position and intended movement of a ship about the navigational characteristics are displayed on ECDIS [7]. An ECDIS uses GPS,

Fig. 1.10. Marine radar

satellite position data, radar, Automatic Identification System (AIS), sensors, depth sounder, NAVTEX, and other data with the complex database, which contains chart decision-making information [8]. Navigation Teletex (NAVTEX) refers to the automatic service of navigation and meteorological warnings and forecasts.

AIS is an important navigation auxiliary system. It uses the technologies, including time-division multiple access (TDMA) and very high frequency (VHF) bands to automatically broadcast and receive static, dynamic, navigation, and safety information of ships and surrounding ships.

Fig. 1.11.   ECDIS

An AIS can implement the identification, monitoring, and communication of ships, using a variety of core technologies, including satellite, digital, information processing, and communication technologies, as shown in Figure 1.12. The AIS interface circuit receives location information from a GNSS receiver, the information of speed from a log, and the course information from a gyrocompass. The processor processes information and sends the static and dynamic information of ship navigation through a VHF transceiver and displays the received navigation data around ships on an ECDIS monitor.

Figure 1.13 shows a data recorder of ship voyage which is a real-time recoverable and safety device to record the location, physical condition, dynamics information, handing data, and navigation data of a ship during an accident. It consists of a processor, sensor interface, microphone set, signal processing circuit, alarm indicator, data protection module, data playback equipment, and power supply.

Fig. 1.12.   AIS

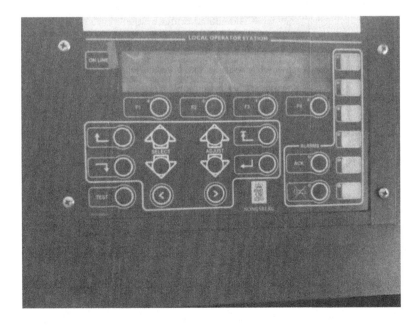

Fig. 1.13.   Voyage data recorder

Fig. 1.14. Integrated bridge system

Integrated Bridge System (IBS) integrates the above navigation instruments, integrating radar, ECDIS, AIS, autopilot, and different types of navigation and ship handling equipment based on Integrated Navigation System (INS), which integrates ship handling, navigation, integrated information display, automatic collision avoidance, navigation management control functions, and communication as shown in Figure 1.14.

1.5. Conclusion

In this chapter, we provide a general overview of a ship and introduce the basic structure, components, and corresponding functions. It is a tutorial for readers who are not in the shipping fields so that they can easily understand the rest of the book.

Acknowledgement

Jia Wang and Tieshan Li's work was supported in part by the National Natural Science Foundation of China (under Grant Nos. 51939001, 61976033); the Science and Technology Innovation Funds of Dalian (under Grant No. 2018J11CY022); the Liaoning Revitalization Talents Program (under Grant Nos. XLYC1908018, XLYC1807046); the Fundamental Research Funds for the Central Universities (under Grant No. 3132019345).

References

1. Y. X. Jin and S. C. Wu, *Ship Structure and Equipment.* China Communications Press, 2012.
2. Z. J. Guan and T. Liu, *Navigation instrument (Book 1: navigation equipment for ships).* Dalian: Dalian Maritime University Press, 2009.
3. J. Wang, Y. Xiao, T. Li, and C. L. P. Chen, "A survey of technologies for unmanned merchant ships," *IEEE Access*, vol. 8, pp. 224 461–224 486, 2020.
4. W. Li, *Ship Structure and Equipment.* Dalian: Dalian Maritime University Press, Aug. 2008.
5. Boe Marine, "Do you need radar?" https://www.boemarine.com/blog/post/do-you-need-radar/, January 2018, accessed Dec. 13, 2018.
6. I. H. Organization, "Electronic navigational charts (encs) & electronic chart display and information system (ecdis)," 2018, accessed March 4, 2019. [Online]. Available: http://www.iho.int/srv1/index.php?option=com_content&view=category&id=72&lang=en&Itemid=287
7. Wikipedia, "Unmanned Surface Vehicle," https://en.wikipedia.org/wiki/Unmanned_surface_vehicle, Accessed April 4, 2019.
8. D. J. Wright and D. J. Barlett, *Marine and Coastal Geographical Information Systems.* CRC press, 2000.

Chapter 2

General Overview of Ships

Hasan Bora Usluer

T.R. Galatasaray University, Ortakoy, Istanbul, Turkey
(e-mail: hbusluer@gsu.edu.tr).

2.1. Introduction

Maritime exploration and utilization started with people and their endless needs. Throughout history, it was seen that it has played an important role in the establishment of states. Poeple started to use the sea and the boats that were built for feeding purposes first and then for exploring the nearby geographies. Over time, people have revealed various types of floating elements, by starting from the material they had and by trying to cope with the concrete problems that the sea beside people. It is well known that three-quarters of the world consists of water which includes oceans, seas, lakes, rivers, etc., as shown in Fig. 2.1. Therefore, humankind had to use water with different kinds of floating devices.

It is said that maritime exploration and later maritime transportation that has been carried out since prehistoric times started with people passing over rivers [2]. Civilization progressed with the development of technology, which caused people to travel by sea using new techniques and those from the old and known land travel. Due to this development, fishing, trade, and transport became easier. The common name of uninterrupted water bodies collected in large basins on earth is the ocean and the sea. Oceans and seas cover 362 million km^2 of the earth's surface (more than 70%). Maritime issues concern the lives of all countries of the world, including those with and without seas, as a state and nation. One of the most prominent examples of this issue is that some states that want to benefit from maritime access indirectly, even though they do not have any shore, engaged in maritime trade activities under the banner of other nations. In other words, they tried

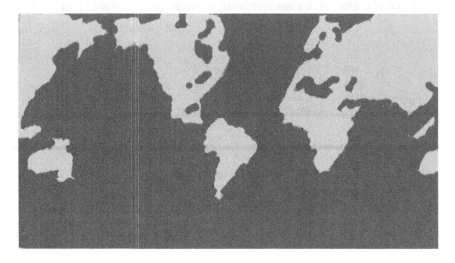

Fig. 2.1.   World ocean map [1]

to create a maritime power without their seas. The seas have been a great field of struggle throughout history, as they have created great potential for national powers. For this reason, Barbaros Hayreddin Pasha (1478-1546), one of the most unforgettable Turkish admiral of world maritime history, once said, He who rules the sea rules the world.

2.2. General History of Ships

As mankind's discovery and understanding of itself grew, so did its understanding of what it can accomplish. People experimented with all kinds of materials that were capable of floating. Particularly noteworthy at this stage is bamboo, logs, reeds, or leather. The first representation of the sailboat in Kuwait, seen in an inscription of a painted disc, was believed to have survived from 5500 to 5000 BC [3]. The produced sailboat offered a chance to sell and teach other civilizations how to build, sail, and navigate ships [4]. In Azerbaijan, the Pesse Canoe, the oldest boat, shown in Fig. 2.2, which was known to be about 3 meters long, had 20 rowers and was discovered 8000BC, is an important example [5]. Some sources write about boats that were produced more than 700,000 years ago. The dugout type of the first known boat is also known as logboat or monoxylon. In general, it was used for fishing and sightseeing purposes by many nations. They had been and were frequently seen in Asia, Africa, Europe, and America

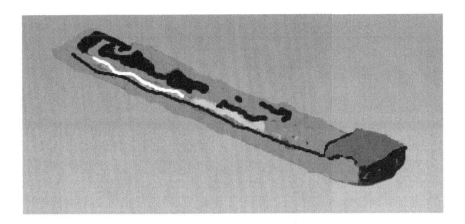

Fig. 2.2.   Azerbaijan's Pesse Canoe, the oldest boat [1]

continents, as well as the Pacific islands, New Zealand, and Australia (by the Aborigines). Another type encountered as a historical boat type is the outrigger, which was known that people were generally used in South and Southeast Asia.

The chart in Fig. 2.3 shows the seaborne migration of the Austronesians beginning at around 3000 BC. While humans were using boats on the seas/rivers, they have learned the sea without knowing the effect of currents and started using marine science.

According to Souza [7], in the old known history of humankind, maritime science is a subject that was discussed and researched by historical

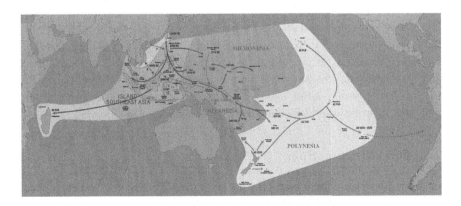

Fig. 2.3.   Seaborne migration of the Austronesians [6]

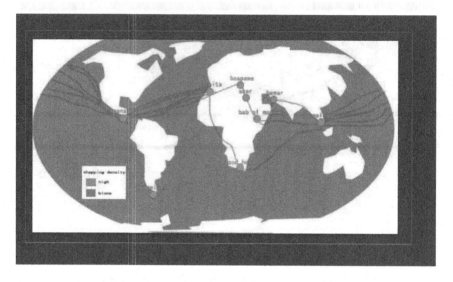

Fig. 2.4.   Important seaways and canals of the world maritime transportation [9]

scientists and archaeologists, as they were riding a horse with a floating log
or similar materials like a buoy [7]. Most of the earth's surface consists of
water, just like the human body. Mankind continued to carry all the materials, starting with getting food from the sea, and continuing with discoveries
and meeting their needs, as well as transportation. Without mankind, ships
and ports would not exist. Besides, there is no world without oceans, an
ocean without a coast, a coast without a harbor, a harbor without ships,
and a sea without people who do not use all of these. Parallel to the developments of the maritime industry and maritime transportation, especially
over the oceans, seas, and straits (natural and artificial), canals and harbors
have disadvantages of increasing sailing safety, regulation of maritime traffic
and safe navigation, as a result of the increase in ship types, numbers, sizes,
and their capacity. In maritime areas with large traffic volumes shown in
Fig. 2.4, increasing sailing safety measures has been made compulsory in
line with international decisions taken by relevant authorities [8].

2.3. Direction of a Ship

In maritime terminology, many words are not understood by everyone but
mariners. As seen below in Fig. 2.5, the left side is port, the right side
is starboard, the center of the shipping line is amidship, the front side is

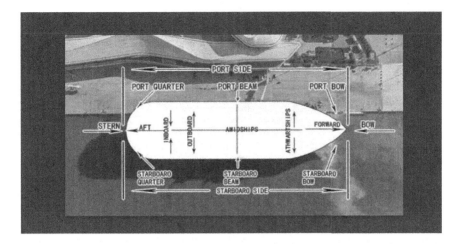

Fig. 2.5.   Directions of the ship [9]

forward, and also the backside is aft [10]. But small differences have existed between forward and aft due to point of view and looking directions. From Fig. 2.5, the front side is forward if the direction from the ship is toward the sea; otherwise, it uses with bow. Also, the backside is aft of the ship if the direction from the ship is toward the sea; otherwise, it is stern.

2.4. Ship Types

Ships diversified according to their needs and usage patterns. The ships, which were originally used for discoveries and expected to be durable for self-defense, started to be used for war. Many factors are of prime importance in designing ships. Early on, the only need was an exploration of the surface. As the transcontinental voyage and coasts were discovered, the needs of maritime nations changed. Meanwhile, the seas and oceans navigated by seafarers forced the ships to change their qualities due to the difficulties caused by meteorological conditions and problems encountered on the discovered shores. In coastal navigation, boats on the raft and similar coasts were used without any damage, but they could not be used in the open sea and difficult sea conditions. Understanding that different products are grown and used in every continent and climate as a result of discoveries, mankind has also developed sea trade due to seaway transportation.

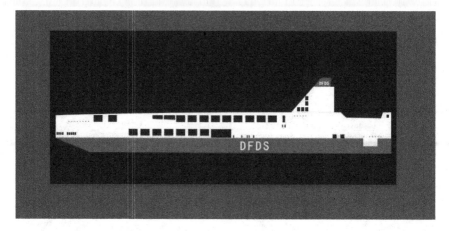

Fig. 2.6. The largest RORO ship is 237 meters long, named Ephesus Seaways, and was berthed Istanbul in 2019

2.4.1. *Cargo Ship*

A cargo ship or freighter is a kind of merchant navy ship that carries cargo from one port, pier, wharf or base to another. Cargo ship is the general name given to the ships used for the transportation of cargo in military, commercial and public services. Loads and modes of transport vary according to needs. Therefore, the types and shapes of ships vary. Ships carry many kinds of cargo and goods through the oceans, from one port to another. Therefore, maritime trade is managing and transporting world international trade.

2.4.2. *RO-RO*

RO-RO stands for Roll on Roll, as shown in Fig. 2.6. RO-RO ship are carriers that are constructed for wheeled cargo such as automobiles, lorries, trailers, etc.

2.4.3. *Cruise Ship*

Cruise transportation is the most important platform that provides the opportunity to have a holiday on luxury ships in addition to sea transportation (see Fig. 2.7). Also, it ensures that international ports are open to tourists and contribute to global trade. Because some ships carry vehicles, both human and vehicle transportation affects maritime trade positively.

Fig. 2.7. Diamond Princess is the largest cruise vessel which was made popular in March 2020 by the COVID-19 pandemic

2.4.4. *Naval Ship*

Naval ships are elements of war belonging to the naval forces working under the armed forces to protect the territorial waters of seafaring countries. Warships or naval ships are very different from civilian commercial ships. The difference in purpose and usage starts with the manufacturing phase, where ships are divided into surface and underwater vessels. Those used underwater are called submarines. Surface ships vary in types, lengths, weapons, and intended use. Aircraft carrier, destroyer, frigate, corvette, patrol boat, fast attack boat, etc. are great examples of surface ships, as shown in Fig. 2.8.

2.4.5. *Tanker Ship*

Tanker ships are types designed to carry chemical and dangerous cargo, such as oil and gas, and are divided according to their content and danger class. The International Convention for the Prevention of Pollution from Ships (MARPOL) is an international marine convention , which was signed and accepted in 1973 and changed in 1978, MARPOL73/78, (see Fig. 2.9). MARPOL has 6 annexes. Annex 1, 2. and 3 concern the transportation of hazardous cargo by tankers. Annex 2 (Regulations for the Control of Pollution by Noxious Liquid Substances in Bulk) gives exact information about chemical tankers and their safe shipping [11].

Fig. 2.8.   Various warships sail together

Fig. 2.9.   M/T Torrey Canyon disaster was a major issue of MARPOL73/78.

2.4.6. *Fishing Vessel*

Fishing and related activities go back to the oldest sea use. Fishing has
been a source of income and economic activity, as it provides a source
of income for people [12, 13]. Vessel operations are included in fishing,

Fig. 2.10. Fishing harbor operation [15].

hunting, transporting, and storing fish. The best known fishing boats are commercial, artisanal, and recreational. According to the United Nations Food and Agriculture Organization (FOA), the most well-known fishing and used boat model in the world is the trawler [14]. Fig. 2.10 shows an example of a fishing harbor operation [15].

2.4.7. *Container Ship*

With the developing technology, transportation types also changed. The need arose to move the loads that can be transported in bulk while preventing damage. Resistant containers were used in both sea and land transportation and were of two types, 20 and 40 feet. Container ships carry cargo loaded in metal containers by trucks and similar vehicles intermodally [16]. Intermodal transportation has standardization. TEU, which stands for twenty-foot equivalent units, is the most significant unit of measurement in container transportation. This measure gained a globally accepted feature by the International Organization for Standardization (ISO). Fig. 2.11 shows the largest container ship of the world, MV HMM Algeciras, which is 399.9 meters tall.

2.4.8. *Bulk Carrier*

Bulk cargo carriers, as illustrated in Fig. 2.12, are the most important kind of carriers in sea trade. Bulk carriers are designed to carry loads such as

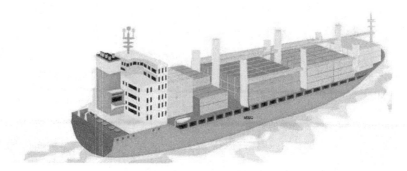

Fig. 2.11.   The largest container ship in the world, MV HMM Algeciras, which is 399.9 meters tall

unpackaged coal, grains, ore, steel coils, and cement. Due to the type of seas used in world maritime transportation, in terms of artificial and natural waterways, sizes and types of ships vary accordingly. The vessel size group depends on deadweight tonnage (DWT) capacity which is a measurement of a ship's weight carrying capacity.

Fig. 2.12.   Bulk carrier during voyage [17]

Fig. 2.13. Reefer ship at harbor [19]

2.4.9. *Reefer Ship*

Fresh fruits and vegetables, meat, fish, packaged dairy products should be transported in cold environments. Reefer ships offer this kind of temperature-controlled transportation , as they are equipped with a refrigerator facility in the hold. Reefers are categorized into three types: conventional, side-door, and containerized [18]. Fig. 2.13 shows a reefer ship at harbor [19].

2.4.10. *Other Types of Ships*

Other types of ships include education/training ships, hydrographic survey ships, icebreakers, pilot boats, lightships, tugboats, push tugs, rescue boats, hovercrafts, fireboats, research ships, floating hospitals, hotels, dredges ships or platforms, floating docks, floating cranes, energy ships, and petroleum research platforms.

2.5. Ship Size

Freighters are generally categorized according to loading capacity, weight (deadweight tonnage or DWT), and usage. Minimum and maximum dimensions such as length and width also show the seaways and harbors where a ship can sail. Draft or draught, which is the minimum depth of water a ship can safely navigate, is a limitation for navigation areas such as straits, canals, and also harbors. The ship has three meaningful dimensions: length, beam, and depth. Length is the distance of the ship from bow to stern. Beam is the weight distance which is from port to starboard side. Depth is a distance measured from keel to deck on the amidship line consisting of two parts: draught and freeboard. Draught is the distance from the keel to the waterline of the ship. Freeboard is the distance from the draught to the deck. Fig. 2.14 shows an example of ship dimensions [9].

Tonnage on ships can be classified as lightweight or deadweight. Lightweight also belongs to the body and machine formed by the construction of the ship. Also, it includes lubrication, boiling, and cooling systems. According to the international maritime organization, SOLAS, the lightweight/lightship or displacement concept is the tonnage value of

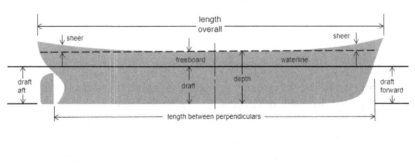

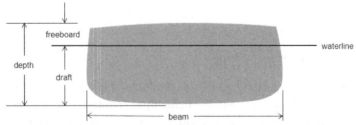

Fig. 2.14. Ship dimensions [9]

weight measured in warehouses and tanks without load, fuel, lubricating oil, ballast water, freshwater, food, crew, passengers, and other items related to them [20]. Deadweight is the tonnage of the weight of all cargo including cargo, crew, passenger, stores, ballast water, freshwater, and sludge. According to SOLAS, deadweight is the difference in tonnes between the displacement of a ship in water of specific gravity of 1.025 at the draught corresponding to the assigned summer freeboard and the lightweight of the ship [21].

2.5.1. *Handysize Ships*

Handysize ships carry dry bulk and oil goods. Handysize ships have a capacity between 10,000 and 35,000 DWT (max 39.999 DWT). Handysize ships are ideal for small harbors with length and draft restrictions. Handysize ships' equipment allows ships to provide easy service in ports lacking loading operation infrastructure.

2.5.2. *Handymax and Supramax Ships*

Handymax and supramax ships also carry dry bulk and oil goods. The capacity of an average handymax ship is usually between 35,000 and 50,000 DWT, while the supramax ship capacity is between 50,000 and 60,000 DWT [22].

2.5.3. *Panamax Ships*

As it is well understood from its name, this type of ships is designed for passage through the Panama canal which connects the Atlantic and Pacific Oceans and it approximately 80 km long. The canal was cut through one of the narrowest saddles of the isthmus that joins North and South America [23]. It is very important, as it provides access to many global sea trade routes easily passing through the canal. This type of Panamax ships is middle sized carriers and eligible for passing through the canal which length is 320.04 meters, width is 33.53 meters, and depth is 12.56 meters [24]. The carrying capacity of Panamax ships is in the range of 60,000 to 79,999 DWT. As a result of the changing requirements of developing technology and maritime trade, new Panamax or post-Panamax sizes announced for the canal in 2009 by the Panama Canal Authority. New Panamax vessel's length changed from 294.13 meters to 366 meters, width from 32.31 meters to 49 meters, and draught from 12.04 meters to 15.2 meters.

2.5.4. *Capesize Ships*

Capesize ships' capacities are between 110,000 to 199,000 DWT. This type of ships is also named as large volume bulker and tankers. Often ships of this size can only approach the largest ports that serve global maritime trade which are designed to accommodate them. Capesize ships are larger than Panamax and Suezmax ships in terms of draft and DWT, while they are smaller than very large crude carriers (VLCC), ultra large crude carriers (ULCC), and bulk ships [25]. The reason for the use of capesize ships is that the draft value of the ship is also eligible to pass via canals. So Capesize ships can use a transitway via Cape Horn to connect the Atlantic and Pacific. As a result of the developing technology and maritime transport requirements, in 2009, the depth of the Suez Canal deepened from 18 meters to 20 meters, and many capesize ship channels began to be used.

2.5.5. *Suezmax Ships*

Suezmax ships get their name from the Suez Canal. This kind of ship is medium- to large-sized. Its tonnage is limited from 120,000 DWT to 200,000 DWT. According to the Suez Canal Authority, the Suez Canal, or Qanat as-Suways in Arabic, is not natural. Its artificial structure reaches from north to south via the Isthmus of Suez in Egypt and connects between the Mediterranean Sea and the Red Sea [26].

2.5.6. *Aframax Ships*

An Aframax ship is a medium-sized crude oil tanker. Its tonnage ranges from 80,000 DWT to 120,000 DWT. AFRA means Average Freight Rate Assessment.

2.5.7. *VLCC and ULCC*

Very large crude carriers (VLCCs) and ultra large crude carriers (ULCCs) ships are also the largest liquid carriers in the world. VLCC is one of the largest working cargo ships in the world with a load-carrying capacity exceeding 250,000 DWT [27]. This kind of carrier is known as supertankers. They are also the preferable ship of all oil major companies. Theu carry out long-range sea transportation from the Persian Gulf to Europe, Asia, and North America [28].

2.5.8. *Q-Max Ships*

Q-Max is an acronym for Qatar Maximum. These kinds of ships are for transporting huge volumes of liquefied natural gas (LNG).Q-Max gas tankers' length is more than 340 meters, width is 53.8 meters, and height is 34.7 meters. The draught of the ship is 12 meters. Their cargo capacity is 266,000 cubic meters of liquefied petroleum gas (LPG), which is equal to 161,994,000 cubic meters of LNG [29].

2.6. Future Ship Concepts

Ships have been shaped according to maritime transportation needs throughout history. However, technology had a positive effect reaching the point of unmanned ships. According to [30], unmanned ships are considered an important element of a competitive and sustainable maritime industry in the future, although they will remote controlled. Industry 4.0 and its effects also influence all maritime sections in the form of Maritime 4.0 and in particular the components of Ship Management 4.0, developments, and requirements related to them [8]. Also, attention will be paid to the changes that will be caused by developments in Ship Management 4.0 in the conventional maritime sector [31]. This is evident in Rolls-Royce and Finferries launching the world's first fully autonomous ferry in maritime transport. Also, the first unmanned autonomous ship sailed between Parainen and Nauvo in December 2018. Finally, the Finland Finferries car carrier "Falco" was controlled with a remote during voyage without bridge personnel [32].

2.7. Conclusion

Maritime transportation has emerged and developed with the existence and needs of human beings. While the seas were used for feeding purposes, they were further developed by the naval forces of various countries and reached the highest level with trade transportation. Three-quarters of the earth is covered with oceans, seas, rivers, and lakes. Thus, both marine and maritime concepts are accepted and used by human being. Endless human needs are the subject of economics. The most appropriate, safe, fast, and inexpensive transportation method of goods that meets human needs is sea transportation. Maritime transportation has served those needs from the past to the present without losing its importance. It will continue to serve with the same importance in the future. The most important elements of developing shipping are ships, ports/harbors, and shipping lines. Merchant navy has been designed using true organization and foundation. This is based on cooperation between operators of both on-board and shore companies, in addition to using the right ship and lines. While the developing technology affects ships in a good way, unmanned ship technology has also developed and has been accepted by international organizations as a result of adverse events such as piracy in sea transportation.

References

1. K. Demircan, "Ocean-map," https://khosann.com/okyanuslar-hakkinda-y anitini-bilmedigimiz-7-soru/ocean-map/, Apr. 2019, accessed February 7, 2021.
2. B. Baki, *Lojistik yönetimi ve lojistik sektör analizi.* Volkan matbaacılık, 2004.
3. Wikipedia, "Sailing," https://en.wikipedia.org/wiki/Sailing, accessed February 7, 2021.
4. R. A. Carter, "Boat remains and maritime trade in the persian gulf during sixth and fifth millennia bc." *Antiquity.*, vol. 80, no. 307, pp. 52–63, 2006.
5. J. Vaucher, "Prehistoric craft," http://www-labs.iro.umontreal.ca/vaucher /History/Prehistoric_Craft/, Apr. 2014, accessed February 7, 2021.
6. Wikipedia, "Austronesian people across the pacific," Tech. Rep., accessed February 7, 2021. [Online]. Available: https://en.wikipedia.org/wiki/Mari time_history
7. P. De Souza, *Seafaring and Civilization: Maritime Perspectives on World History.* Profile, 2002.
8. H. Bora, "Gem trafk hzmetlernde denz hartaciliinin nem. usluer," http://ww w.vts.org.tr/i-ulusal-gemi-trafik-hizmetleri-kongresi/, 2014, accessed February 7, 2021.
9. R. Asariotis, H. Benamara, J. Lavelle, and A. Premti, "Maritime piracy. part i: An overview of trends, costs and trade-related implications," 2014.
10. N. O. Service, "Why do ships use port and starboard instead of left and right?" https://oceanservice.noaa.gov/facts/port-starboard.html, 2014, accessed February 7, 2021.
11. IMO, "Marpol," http://www.imo.org/en/About/Conventions/ListOfConve ntions/Pages/International-Convention-for -the-Prevention-of-Pollution-from-Ships-(MARPOL).aspx, accessed February 7, 2021.
12. M. Nuñez-Sanchez, L. Perez-Rojas, L. Sciberras, and J. R. Silva, "Grounds for a safety level approach in the development of long-lasting regulations based on costs to reduce fatalities for sustaining industrial fishing vessel fleets," *Marine Policy*, vol. 113, p. 103806, 2020.

13. F. Uğurlu, S. Yıldız, M. Boran, Ö. Uğurlu, and J. Wang, "Analysis of fishing vessel accidents with bayesian network and chi-square methods," *Ocean Engineering*, vol. 198, p. 106956, 2020.
14. F. . A. Organization, "Fishing vessel types. trawlers. technology fact sheets." http://www.fao.org, accessed February 7, 2021.
15. N. herald, "Japan resumes commercial whaling. but is there an appetite for it?" https://www.nzherald.co.nz/business/japan-resumes-commercial-wha ling-but-\\is-there-an-appetite-for-i t/3TZHBNJUL3N5V4COACDL745KZM/, Jul. 2019, accessed February 7, 2021.
16. Wikipedia, "Container_ship," Tech. Rep., 2020, accessed February 7, 2021. [Online]. Available: https://en.wikipedia.org/wiki/Container_ship
17. Blogspot, "bulk-carriers," http://freight-charter.blogspot.com/2015/04/ty pes-of-bulk-carriers.html, Apr. 2015, accessed February 7, 2021.
18. P. Kohli, "Reefer vessels an introduction," http://www.crosstree.info/Doc uments/reefer%20vessels.pdf, accessed February 7, 2021.
19. T. Taro Lennerfors and P. Birch, *Snow in the Tropics: A History of the Independent Reefer Operators.* Brill, 2019.
20. IMO, "Solas regulations,definitons,4-28," https://www.samgongustofa.is /media/english/SOLAS-Consolidated-Edition-2018.docx.pdf, 2018, accessed February 7, 2021.
21. ——, "Solas regulations,definitons,4-28," https://www.samgongustofa.is/m edia/english/SOLAS-Consolidated-Edition-2018.docx.pdf.docx.pdf, 2018, accessed February 7, 2021.
22. M. Connector, "Handymax," http://maritime-connector.com/wiki/handym ax/, accessed February 7, 2021.
23. C. de Panama, "This is the canal," https://www.pancanal.com/eng/acp/a si-es-el-canal.html, accessed February 7, 2021.
24. M. connector, "Panamax and new panamax," http://maritime-connector.c om/wiki/panamax/, accessed February 7, 2021.
25. ——, "Capesize," http://maritime-connector.com/wiki/capesize/, accessed February 7, 2021.
26. Suezcanal, "Sca - home," https://www.suezcanal.gov.eg/English/Pages/def ault.aspx, accessed February 7, 2021.
27. F. Cătălin, D. Cosmin-Laurenţiu, B. Nicolae, and S. Liviu-Constantin, "Impact of hvac system upon functioal parameters of main engine for a vlcc ship," *Journal of Marine Technology and Environment Year 2020, Vol.*, vol. 1, p. 11, 2020.
28. M. connector, "Vlcc and ulcc," http://maritime-connector.com/wiki/vlcc/, accessed February 7, 2021.
29. ——, "Q-max," http://maritime-connector.com/wiki/q-max/, accessed February 7, 2021.

30. L. Kretschmann, H.-C. Burmeister, and C. Jahn, "Analyzing the economic benefit of unmanned autonomous ships: An exploratory cost-comparison between an autonomous and a conventional bulk carrier," *Research in transportation business & management*, vol. 25, pp. 76–86, 2017.

31. H. Usluer, "Industry 4.0 and autonomous ships effects on marine environment," bosnia Herzigovina : ICOEST-5th International Conference on Environment Science and Technology, ISBN 978-605-81426-2-6,.

32. Rolls-Royce, "Rolls-royce and finferries demonstrate world's first fully autonomous ferry," https://www.rolls-royce.com/media/press-releases/2018/03-12-2018-rr-and-finferries-demonstra te-worlds-first-fully-autonomous-ferry.aspx, Dec. 2018, accessed February 7, 2021.

PART 2

Formation Control, Artificial Intelligence, and Simulation in Smart Ships

Chapter 3

Unmanned Ship Formation

Yue Yang[1], Yang Xiao[2,*], Tieshan Li[3,1]

[1] *Navigation College, Dalian Maritime University, Dalian, Liaoning 116026, China (e-mail: yueyang@ieee.org).*

[2] *Department of Computer Science, The University of Alabama, Tuscaloosa, AL 35487-0290, USA (e-mail: yangxiao@ieee.org).*
* Yang Xiao is the corresponding author.

[3] *School of Automation Engineering, University of Electronic Science and Technology of China, Chengdu, 611731, China (e-mail: tieshanli@126.com).*

3.1. Introduction

Seventy-one percent of the earth is covered by oceans, and oceans have vast resources [1]. With the depletion of mineral resources (oil, coal, natural gas, etc.) on the land, the exploration and utilization of ocean resources are accelerated. Human activities in the ocean increase sharply and cause serious damage to ocean environments. Moreover, due to the harsh ocean environments, human safety and work efficiency are difficult to guarantee when performing a task. Therefore, unmanned ships attract a lot of attention, as they can greatly improve the efficiency of ocean exploration and ensure human safety. However, in a complex marine environment, a single unmanned ship is limited by its ability and difficult to complete complex tasks.

Inspired by the cooperative behaviors in nature, the theory of multi-ship cooperative control has been developed rapidly. In nature, cooperative behaviors of biological groups are ubiquitous [2], such as cooperative transport of colony ants and avoidance of predators by groups of fish [3]. Cooperative behaviors play a key role in the group feeding and defending against natural enemies. If cooperative behaviors are applied to a ship swarm, many ships can achieve complex swarm behaviors. Cooperative behaviors can greatly overcome the limitation of a single ship's capability. Compared to a single ship, cooperative multi-ships have better robustness, flexibility, and higher operating efficiency.

Formation control is a significant topic in cooperative control fields [4]. Formation control is defined as a group of agents being controlled to maintain predefined spatial patterns while performing a specific task with the desired trajectory. Ship formation becomes one of the most attractive research issues in both academic and engineering fields since ship formation has various applications. In search and rescue fields, ship formation can increase search area and improve the rescue efficiency by forming a line formation. In military fields, ship formation can achieve escort missions, formation patrol, battle attack/defense formation, and so on.

The rest of the chapter is organized as follows. In Section 3.2, we present network topologies of ship formation. In Section 3.3, control structures of ship formation are presented. We introduce control methods of formation and multi-ship formation in Section 3.4 and Section 3.5, respectively. Finally, we conclude this chapter in Section 7.8.

3.2. Network Topologies of Ship Formation

Network topologies refer to layouts of networks. Since network topologies define or describe where different ships are placed and how these surface ships interconnect with each other, designing a suitable network topology is essential to achieve better formation. In some situations, ship formation just needs two or three ships to fulfill tasks. However, for some tasks (e.g., searching for a sunken ship), ship formation needs a large scale. Most network topologies can be classified into centralized, distributed, and cluster topologies.

3.2.1. *Centralized Topologies*

In a centralized network scenario, communications between surface ships take place through a central station, such as a leader ship. The central station sends information to peripheral surface ships, whereas peripheral ships do not need to communicate with each other. Similarly, a peripheral ship can also work as a sub-central station with its sub-peripheral ships. An

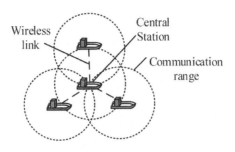

Fig. 3.1.   Star topology

obvious disadvantage of centralized topologies is that if a central station or the connection to the central station fails, the peripheral ships of the central station fail to communicate. Centralized topologies include star topologies and tree topologies.

As shown in Fig. 3.1, in a star topology, a central station can communicate with peripheral ships, whereas peripheral ships do not need to communicate with each other. Star topologies are easy to design, but formation scales are limited. Limited scales defined as formation scales are determined by communication ranges between the central station and each peripheral ship since no ship works as a relay node in star topologies.

Tree topologies are hierarchical topology structures shown in Fig. 3.2. Similar to star topologies, a central station in a tree topology has many peripheral surface ships. Moreover, some of these peripheral ships can also

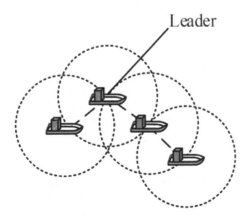

Fig. 3.2.   Tree topology

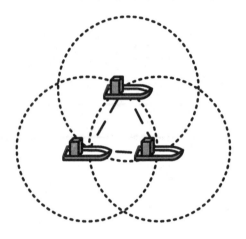

Fig. 3.3.   One-hop topology

have their sub-peripheral ships. One advantage of tree topologies is that topology robustness can be improved by distributing the functions of the central station. Another advantage is that tree topologies can scale up ship formation since formation can be extended by placing ships as relay nodes.

3.2.2. *Distributed Topologies*

In contrast to centralized topologies, distributed topologies do not have a central station. Ships can directly communicate with one or more ships. Distributed topologies include one-hop topologies and multi-hop topologies.

As shown in Fig. 3.3, in a one-hop topology, distances between ships are small enough that each ship can communicate with its neighbors directly. The failure of any single ship will not cause the entire formation to fail.

Figure 3.4 shows a multi-hop topology in which transmitters send information to receivers through one or more relay nodes. Multi-hop topologies can have large coverage, whereas communication delay increases with the increasing number of relay nodes.

3.2.3. *Cluster Topologies*

Cluster topologies combine centralized and distributed topologies, as shown in Fig. 3.5. The whole ship formation is divided into several clusters and each cluster has its ship leader. Cluster leaders communicate with each other through a distributed topology. The inside of each cluster implements

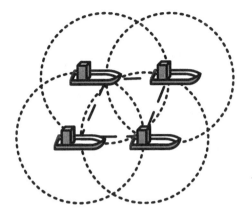

Fig. 3.4. Multi-hop topology

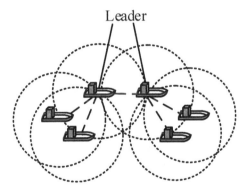

Fig. 3.5. Cluster topology

a centralized topology. Cluster topologies have better scalability and robustness for ship formation. For robustness, if one of the leaders fails, the
other clusters still maintain formation well.

3.3. Control Structures of Ship Formation

Network topologies are used to describe relative locations and communication relations of ships. Control structures are used to describe the difference
in the decision-making process. To date, control structures can be classified
into three types: centralized, decentralized, and distributed structures.

Since centralized structures control the group uniformly through the master control unit, they must collect all environmental and individual information. Decentralized structures do not have a master control unit. In decentralized structures, each control unit is equal and only has a part of the group information. Although decentralized structures are difficult to ensure the global optimal control with local information, they have several advantages, such as fault tolerance and reliability. In distributed structures, a single agent is controlled based on local information, and each control unit also does not need to have global information states of the group. Distributed structures are flexible and have good fault tolerance and extendibility.

3.4. Control Methods of Formation

Control methods can be classified into three categories: leader-follower, virtual-structure, and behavior-based.

A basic idea of leader-follower methods is that one or more agents in a group are designated as leaders, and the other agents are designated as followers [5,6]. The followers track the leader's position at the desired distance and achieve formation movement with the leader. The authors in [7], for the first time, proposed a leader-follower method to realize formation control of mobile robots with two kinds of formation controllers: $l - l$ and $l - \theta$. The goal of the $l - l$ controllers is to keep the followers at the desired distance from two leaders. The goal of $l - \theta$ controller is to keep the followers at the desired distance from a leader at the desired angle. The authors in [8] study vision-based leader-follower formation control problems. Taking into account the three situations in which the leader's velocity and acceleration are known, the rate of change of position is known, and the velocity and acceleration are unknown, the authors in [8] propose three kinds of leader-follower formation control algorithms, including full state, robust state, and output feedback algorithms. The authors in [9] study the leader-follower formation control with input constraints. For leader-follower methods, since the leader's state of motion is often uncertain, followers also have complex dynamic characteristics. Therefore, the robustness of formation controllers attracts much attention. The authors in [10] achieve leader-follower formation control by only using position information feedback. The authors in [10] estimate the time-invariant or time-varying velocity of the leader and propose an adaptive formation control algorithm. Based on feedback linearization and sliding mode control techniques, the authors in [11] propose a robust adaptive controller that takes into account the uncertainty of system parameters. The proposed control algorithm in [11] is robust to the uncertainty of leader motion. Unlike the above papers, which only consider kinematic models of controlled objects, the authors in [12] study the formation control of mobile robots under the dynamic models. In [12], the authors adopt back-stepping methods to design a controller. In [12], uncertain dynamics of followers are compensated by a neural network. Moreover, the authors in [13] utilize a neural network and feedback of robust integral sign errors to achieve asymptotic tracking effects. The methods in [12] and [13] require the follower speed state to be measurable, and this requires the actual use of speed sensors and increases equipment costs.

The advantages of leader-follower methods are simple and easy to implement since controllers can achieve formation by decreasing tracking errors between leaders and followers. A disadvantage is that leader-follower controllers lack formation feedback, leading to instability of the entire formation when a leader fails. Another disadvantage is the superposition effect of formation tracking errors and is discussed in [14]. Leader-follower methods simplify formation control issues into several independent tracking problems, and each follower only needs to obtain the state information of the leader. Therefore, most leader-follower methods belong to decentralized structures. In recent years, with the development of multi-agent control technologies, leader-follower methods are promoted from control methods to a controlling idea and are widely concerned with the coordinated formation control of multi-agent systems. In multi-agent systems, leaders are the tracking targets or references for the entire system and they can be real or virtual.

For virtual-structure methods, the whole formation structure is designed as a rigid body and every individual chooses different reference points on the rigid body as their tracking targets. When the rigid body moves, individuals can achieve a fixed formation by following reference points on the rigid body. Designing a virtual-structure method has three steps: (1) defining desired dynamic characteristics of the whole virtual structure; (2) transforming virtual-structure motion into desired individual motions; and (3) obtaining trajectory tracking control methods of formation individuals. The idea of virtual-structure methods is proposed for the first time by the authors in [15]. Inspired by [15], the authors in [16] utilize virtual-structure methods to achieve spacecraft formation. However, the controllers in [15] and [16] lack feedbacks from followers. Once a follower falls behind, the entire formation can fail. Therefore, the authors in [17] define formation errors by a Lyapunov function and improve formation stability through utilizing formation feedback mechanisms. The above papers all adopt centralized structures, and each individual fails can cause an entire formation to fail. Therefore, the distributed virtual-structure methods are studied in [18] and [19].

Advantages of virtual-structure methods include that group movements are easily defined and tracking effects have high precision. Disadvantages of virtual-structure methods include lack of flexibility and adaptability and being only suitable for small group formation.

For behavior-based methods, each behavior and local rule of formation is designed to achieve a group behavior. Each individual has several basic behavior patterns and each behavior pattern has specific goals, including obstacle avoidance, collision avoidance, target tracking, and formation keeping. The authors in [20], for the first time, propose behavior formation and design behavior controllers to keep formation. In the behavior-based method, formation individuals make decisions according to the behaviors of neighboring individuals. Therefore, behavior-based methods belong to decentralized structures.

The disadvantage of behavior-based methods is that they are difficult to analyze mathematically. As a result, the behavior of the formation group is difficult to predict and the stability of the formation is difficult to guarantee. Behavior-based methods are less used in multi-ship formation control.

3.5. Multi-ship Formation

Control issues of single-ship movement can be classified into point stabilization, trajectory tracking, and path following. Since the 1990s, control issues of a single ship have been studied extensively, especially for the nonlinear control of under-actuated ships. With the development of single-ship control technologies, ship movement control and formation control have been integrated and developed together. Next, we discuss the applications of three formation control methods in multi-ship formation control.

Due to easy implementation and simple design, leader-follower methods are widely applied in multi-ship formation. In [21], the authors study under-actuated ship formation with consideration of dynamic model parameter uncertainty and time-varying disturbance. In [21], the authors adopt sliding-mode control and design $l - l$ and $l - \theta$ controllers. Based on sliding-mode control, the authors in [22] propose a parameter estimation method to overcome environment disturbances. In [22], a continuous proportion-integration sliding-mode term is used to eliminate vibration effect of a closed-loop system. With limitations of the line of sight distance and tracking errors, the authors in [23] propose fault-tolerant leader-follower formation to deal with the situations of several actuator failures. To solve tracking errors of ships in sight distance and direction angle, the authors in [24] design a reconstruction module to estimate the velocity vector of the leader and design a controller by using back-stepping methods. Based on the predictive control, the authors in [25] achieve that ships can avoid obstacles and reach a destination in a certain formation shape without environment knowledge and a predefined trajectory.

In fields of multi-ship formation, virtual-structure methods can be transformed into two control issues: coordinated path tracking and coordinated target tracking. Coordinated path tracking means that ships in a fleet need to track a parameterized predetermined trajectory. Coordinated target tracking means that ships in a fleet need to track a reference point or a target point. Coordinated path tracking is studied by the authors in [26] for the first time in ship formation fields. In [26], the authors predefine reference points of virtual structures for each ship and achieve ship formation based on kinematics and dynamics control. The kinematic control makes the ship converge to a specified formation reference path and the dynamic control makes the ship move along the path at a specified speed. Based on [26], the authors in [27] adopt passivity-based theories to eliminate the requirement of global communication in [26] and overcome problems of communication bandwidth constraints. Moreover, the authors in [28] further propose a theoretical framework of coordinated path tracking to realize ship formation. The framework in [28] includes two layers: a coordination layer and a control layer. The control layer implements path-following algorithms, and the coordination layer implements consensus algorithms to achieve speed consensus. Although coordinated path tracking adopts distributed control in the coordination layer, the method is still a centralized control since it depends on the global information. For coordinated target tracking, based on the Lyapunov stability theory and back-stepping methods, the authors in [29] propose a robust adaptive tracking controller. In [29], the controller only relies on the position information of a virtual target.

For behavior-based methods, the authors in [30] utilize null-space-based behavioral control to achieve ship formation. The controller in [30] consists of a null-space vector part and a dynamic control part. Null-space-based behavioral control is a centralized structure since the controller assigns tasks to individuals, such as obstacle avoidance and formation maintenance. The null-space vector part provides reference signals to the dynamic control part. Based on the signals, the dynamic control part control speed and direction of ships. Inspired by Lagrangian mechanics, the authors in [31] design a series of constraint functions to achieve ship formation. The constraint functions determine constraint force and the force is used to maintain formation shape. In [32], the authors study distributed consensus formation control under general integral dynamics and undirected topological networks. The authors in [33] study distributed robust consensus formation of under-actuated ships under uncertain disturbances. In [33], dynamics and kinematics equations of ships are transformed into nonlinear systems. Based on the characteristics of nonlinear systems, the authors in [33] propose a state-based controller to ensure robust consensus formation of ships.

In ship formation fields, researchers mainly focus on leader-follower and virtual-structure methods since the two kinds of methods are easier to achieve formations than behavior-based methods. However, behavior-based methods have the advantage of multi-tasking. In recent studies, more and more behavior-based methods are integrated with other methods to accomplish formation and obstacle avoidance at the same time [34], [35].

3.6. Conclusion

In this chapter, we introduce network topologies and structures of ship formation and provide an overview of main formation control methods. We describe a series of significant researches based on the three kinds of formation control methods.

Acknowledgement

Yue Yang and Tieshan Li's work was partially supported by the National Natural Science Foundation of China (under Grant Nos. 51939001, 61976033); the Science and Technology Innovation Funds of Dalian (under Grant No. 2018J11CY022); the Liaoning Revitalization Talents Program (under Grant Nos. XLYC1908018, XLYC1807046); and the Fundamental Research Funds for the Central Universities (under Grant No. 3132019345).

References

1. J. Wang, Y. Xiao, T. Li, and C. L. P. Chen, "A survey of technologies for unmanned merchant ships," *IEEE Access*, vol. 8, pp. 224 461–224 486, 2020.
2. J. Liu, Y. Xiao, Q. Hao, and K. Ghaboosi, "Bio-inspired visual attention in agile sensing for target detection," *International Journal of Sensor Networks*, vol. 5, pp. 98–111, 2009.
3. Y. Zhang, Y. Xiao, Y. Wang, and P. Mosca, "Bio-inspired patrolling scheme design in wireless and mobile sensor androbot networks," *Wireless Personal Communications*, vol. 92, p. 13031332, 2017.
4. Q. Shi, T. Li, J. Li, C. P. Chen, Y. Xiao, and Q. Shan, "Adaptive leader-following formation control with collision avoidance for a class of second-order nonlinear multi-agent systems," *Neurocomputing*, vol. 350, pp. 282–290, 2019.
5. J. Zheng, Y. Huang, and Y. Xiao, "The effect of leaders on the consistency of group behavior," *International Journal of Sensor Networks*, vol. 11, pp. 126– 135, 2012.
6. J. Zheng, Y. Huang, Y. Wang, and Y. Xiao, "The effectsof wireless communication failures on group behavior of mobile sensors," *Wireless Communications and Mobile Computing*, vol. 14, p. 380395, 2014.
7. P. K. Wang, "Navigation strategies for multiple autonomous mobile robots moving in formation," *Journal of Robotic Systems*, vol. 8, no. 2, pp. 177–195, 1991.
8. O. Orqueda, X. Zhang, and R. Fierro, "An output feedback nonlinear decentralized controller for unmanned vehicle co-ordination," *International Journal of Robust and Nonlinear Control: IFAC-Affiliated Journal*, vol. 17, no. 12, pp. 1106–1128, 2007.
9. L. Consolini, F. Morbidi, D. Prattichizzo, and M. Tosques, "Leader–follower formation control of nonholonomic mobile robots with input constraints," *Automatica*, vol. 44, no. 5, pp. 1343–1349, 2008.
10. K. Choi, S. Yoo, J. B. Park, and Y. Choi, "Adaptive formation control in absence of leader's velocity information," *IET Control Theory & Applications*, vol. 4, no. 4, pp. 521–528, 2010.
11. L. Shi-Cai, T. Da-Long, and L. Guang-Jun, "Robust leader-follower formation control of mobile robots based on a second order kinematics model," *Acta Automatica Sinica*, vol. 33, no. 9, pp. 947–955, 2007.
12. T. Dierks and S. Jagannathan, "Asymptotic adaptive neural network tracking control of nonholonomic mobile robot formations," *Journal of Intelligent and Robotic Systems*, vol. 56, no. 1-2, pp. 153–176, 2009.
13. ——, "Neural network control of mobile robot formations using rise feedback," *IEEE Transactions on Systems, Man, and Cybernetics, Part B (Cybernetics)*, vol. 39, no. 2, pp. 332–347, 2008.
14. H. G. Tanner, G. J. Pappas, and V. Kumar, "Leader-to-formation stability," *IEEE Transactions on Robotics and Automation*, vol. 20, no. 3, pp. 443–455, 2004.

15. M. A. Lewis and K.-H. Tan, "High precision formation control of mobile robots using virtual structures," *Autonomous Robots*, vol. 4, no. 4, pp. 387–403, 1997.
16. R. W. Beard, J. Lawton, and F. Y. Hadaegh, "A coordination architecture for spacecraft formation control," *IEEE Transactions on Control Systems Technology*, vol. 9, no. 6, pp. 777–790, 2001.
17. P. Ogren, M. Egerstedt, and X. Hu, "A control lyapunov function approach to multiagent coordination," *IEEE Transactions on Robotics and Automation*, vol. 18, no. 5, pp. 847–851, 2002.
18. W. Ren and R. W. Beard, "Decentralized scheme for spacecraft formation flying via the virtual structure approach," *Journal of Guidance, Control, and Dynamics*, vol. 27, no. 1, pp. 73–82, 2004.
19. J. R. Lawton, R. W. Beard, and B. J. Young, "A decentralized approach to formation maneuvers," *IEEE Transactions on Robotics and Automation*, vol. 19, no. 6, pp. 933–941, 2003.
20. T. Balch and R. C. Arkin, "Behavior-based formation control for multirobot teams," *IEEE Transactions on Robotics and Automation*, vol. 14, no. 6, pp. 926–939, 1998.
21. F. Fahimi, "Sliding mode formation control for under-actuated autonomous surface vehicles," in *2006 American Control Conference*. Minneapolis, MN, USA: IEEE, Jul. 2006, pp. 4255–4260.
22. Z. Sun, G. Zhang, Y. Lu, and W. Zhang, "Leader-follower formation control of underactuated surface vehicles based on sliding mode control and parameter estimation," *ISA Transactions*, vol. 72, pp. 15–24, 2018.
23. X. Jin, "Fault tolerant finite-time leader–follower formation control for autonomous surface vessels with los range and angle constraints," *Automatica*, vol. 68, pp. 228–236, 2016.
24. J. Ghommam and M. Saad, "Adaptive leader–follower formation control of underactuated surface vessels under asymmetric range and bearing constraints," *IEEE Transactions on Vehicular Technology*, vol. 67, no. 2, pp. 852–865, 2017.
25. X. Sun, G. Wang, Y. Fan, D. Mu, and B. Qiu, "A formation collision avoidance system for unmanned surface vehicles with leader-follower structure," *IEEE Access*, vol. 7, pp. 24 691–24 702, 2019.
26. R. Skjetne, S. Moi, and T. I. Fossen, "Nonlinear formation control of marine craft," in *Proceedings of the 41st IEEE Conference on Decision and Control, 2002*. Las Vegas, NV, USA: IEEE, Dec. 2002, pp. 1699–1704.
27. I.-A. F. Ihle, M. Arcak, and T. I. Fossen, "Passivity-based designs for synchronized path-following," *Automatica*, vol. 43, no. 9, pp. 1508–1518, 2007.
28. R. Ghabcheloo, A. Pascoal, C. Silvestre, and I. Kaminer, "Coordinated path following control of multiple wheeled robots with directed communication links," in *Proceedings of the 44th IEEE Conference on Decision and Control*. Seville, Spain: IEEE, Dec. 2005, pp. 7084–7089.
29. R. Cui, S. S. Ge, B. V. E. How, and Y. S. Choo, "Leader–follower formation control of underactuated autonomous underwater vehicles," *Ocean Engineering*, vol. 37, no. 17-18, pp. 1491–1502, 2010.

30. F. Arrichiello, S. Chiaverini, and T. I. Fossen, "Formation control of underactuated surface vessels using the null-space-based behavioral control," in *2006 IEEE/RSJ International Conference on Intelligent Robots and Systems.* Beijing, China: IEEE, Oct. 2006, pp. 5942–5947.

31. I.-A. F. Ihle, J. Jouffroy, and T. I. Fossen, "Formation control of marine surface craft: A lagrangian approach," *IEEE Journal of Oceanic Engineering*, vol. 31, no. 4, pp. 922–934, 2006.

32. F. Ye, H. Dong, Y. Lu, and W. Zhang, "Consensus controllers for general integrator multi-agent systems: analysis, design and application to autonomous surface vessels," *IET Control Theory & Applications*, vol. 12, no. 5, pp. 669–678, 2018.

33. M. Mirzaei, N. Meskin, and F. Abdollahi, "Robust consensus of autonomous underactuated surface vessels," *IET Control Theory & Applications*, vol. 11, no. 4, pp. 486–494, 2016.

34. F. Fahimi, "Sliding-mode formation control for underactuated surface vessels," *IEEE Transactions on Robotics*, vol. 23, no. 3, pp. 617–622, 2007.

35. K. D. Do, "Synchronization motion tracking control of multiple underactuated ships with collision avoidance," *IEEE Transactions on Industrial Electronics*, vol. 63, no. 5, pp. 2976–2989, 2016.

Chapter 4

AI-based Techniques for Smart Ships

Rachit Garg, Arvind W Kiwelekar, Laxman D Netak

Department of Computer Engineering
Dr. Babasaheb Ambedkar Technological University
Lonere 402103 Dist Raigad (MS) India
(e-mails: rachit.garg.nitttr@gmail.com, awk@dbatu.ac.in,
ldnetak@dbatu.ac.in).

4.1. Introduction

Change is constant in the maritime industry. In the early seventies, the introduction of shipping containers transformed the trade and transportation industry. Shipping containers radically changed how shipping, loading, and unloading of goods are done [1]. The introduction of shipping containers, considered as a historical event, has initiated the process of globalization.

The transportation industry, especially air and surface, has revolutionized itself to leverage the strengths of Information and Communication Technologies (ICT) by embracing the concept of smart vehicles. The maritime industry is currently experiencing a similar revolution triggered by the onslaught of ICT. These technologies amplified the process of digitalization leading to disruptive innovation in the maritime industry, called smart ships.

A smart ship has on board data analytics tools, mechanisms to connect on shore control rooms, and tools to manage maintenance, navigation, and communication operations. Traditional ships are operated by a human from the shore, whereas smart ships are managed by software. Cargo shipbuilders are increasingly adopting Artificial Intelligence (AI) and allied technologies to deliver smart ships. With the maturing of smart ship technology, more ships will likely be switched from a manual mode of operation to an autonomous one.

The emerging concept of smart ship has the potential to impact the overall functioning of the maritime industry. Smart ships have the potential to offer new opportunities to all the stakeholders in the maritime ecosystem. The use of autonomous vessels is expected to bring more effectiveness in the delivery of products and services in the maritime sector.

In this chapter, we describe the technologies that are enabling the design of smart ships. In particular, we describe a subset of technologies called AI and allied technologies. Our objective while describing these technologies is to explain their applications in the shipping industry rather than to explain underlying theoretical concepts, technical strengths, and limitations of these technologies.

The rest of the chapter is organized as follows: Section 2 describes the various design parameters to be considered while constructing a smart ship. Various functions that need to support the smart ship are described in Section 3. Section 4 describes the different levels of autonomy achievable through the use of technologies. Applications of AI and allied technologies to build smart ships are reviewed in Section 5. Section 6 presents cost and benefit analysis of adopting AI technologies.

4.2. Smart Ships Design Parameters

This section reviews some of the high-level design parameters that need to be considered while designing smart ships.

4.2.1. *Basic Elements of Smart Ship*

Smart ships are evolved from manually-operated conventional ships [2–4]. Adopting modern day's technologies to develop a traditional ship results in a smart ship. These advanced technologies provide various ways of connecting a ship to different on-shore and off-shore maritime processes. Hence, these technologies transform a conventional ship into a smart-ship [4]. In general, smart-ships consist of three main elements. As shown in Figure 4.1, these three elements and a physical ship form the backbone of a smart ship.

(1) **Navigation**: The power of data and communication empower navigation facility in smart ships. The objective of navigation is to measure motion parameters like position, course, and speed. Various on board

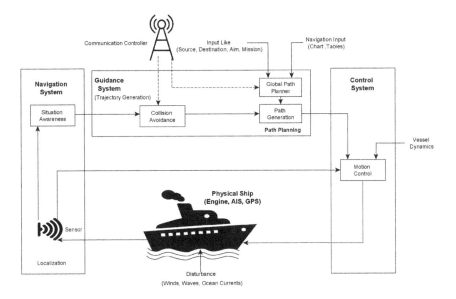

Fig. 4.1.   Basic elements in a smart ship.

sensors collect this information. A software system stores the relevant data, organizes and shares it with other systems and teams. It creates a real-world image to optimize operations and to speed up the decision-making process [5–7].

(2) **Guidance**: The guidance sub system of a ship performs path planning. The guidance sub system uses the real-world image created by the navigation sub system to plan the ship's path. Several parameters need to be considered while planning the path. For example, avoidance of collision, nearby obstacles, and ships, source and destination, the status of other ships are some of the parameters considered for guidance.

(3) **Control**: The control sub system executes the motion control. The purpose is to steer a ship in the right direction. A smart ship has a software-based control system that processes the data. It further issues commands for various control operations: (1) starting, (2) speed control, (3) reversing, (4) stopping, (5) switching off the supply automatically if operating conditions become abnormal, and (6) isolation of motor and control equipment from the supply [8].

4.2.2. *Types of Ships*

The choice of a ship is one of the first design parameters that need to be considered while deciding upon the technologies for smart ships. The type of ship influences various parameters which include the achievement of good results, identification of focus areas, the fulfillment of the demand for new tonnage, and willingness to adjust the existing tonnage. For example, the Rolls Royce focused on a *medium-sized bulk carrier* for constructing smart-ships. This type of ship has a relatively small number of crew members. Furthermore, it has been observed that the cost savings resulting from reduced crew members are modest, for *smaller types of ships*, as compared to the cost savings resulted from autonomy and electrical propulsion.

Some of the smaller types of ships include (i) small or minor island ferries, (ii) tugs, (iii) barges, (iv) supply or service vessels.

4.2.2.1. *Small or Minor Island Ferries*

Small ferries to islands are the lifeline of societies. The costs of operating these ferries are relatively high. To reduce the operating cost, it is normally suggested to combine electrical propulsion with a certain degree of autonomy. Here, *some degree of autonomy* refers to let the ferry crew members deal with the technical and navigational aspects of ferry operations. While, autonomy deals with safety-related aspects of passengers. It is therefore crucial how this role handled if the crew is to be reduced or eliminated from the ferry. An easy solution may be to retain the crew during the day when the number of passengers is projected to be the highest and to reduce or eliminate the crew at times when the number of passengers is small.

4.2.2.2. *Tugs*

Tugs are the vessels that can be autonomous or remotely controlled [9]. This kind of work-boats operates for a limited period in specific geographic areas. For example, RAmora is one such example of a tele-operated work-boat. An experienced tugmaster can remotely control and access workboats in the RAmora series. The remote control and access offer the same experience to a tugmaster, on board controlling of a workboat.

4.2.2.3. *Barges and Lighters*

Barges and lighters are self-propelled boats useful in shorter sea shipping and for the transport of goods and equipment. For example, the *ReVolt* is a modern unmanned and autonomous shipping concept. It provides an efficient solution for increasing transportability requirements. These are the ships that are autonomous, fully battery-driven, and highly efficient in terms of carrying capacity [10].

4.2.2.4. *Supply Vessels*

These are small supply vessels transporting supplies, personnel, and technicians for drilling platforms and offshore wind farms. These vessels can presumably be made autonomous and fitted with electrical propulsion [11]. For example, *Hrnn* is a light, offshore autonomous utility vessel engaged in offshore energy, scientific, hydrographic, and offshore fishing industries [12]. These types of ships are distinguished by their obvious use in the development of autonomous systems as well as electrically powered systems and any other alternative types of energy.

4.2.3. *Safety and Reliability Issues*

The third most important design parameter to be considered while designing smart ships is safety and reliability issues.

The safety of unmanned units implies that autonomous systems must not harm persons or vital equipment. The reduced weight of smart-ships causes less damage in terms of safety as compared to conventional ships. However, it is insufficient to achieve the desired level of safety. Risks involved in ensuring the safety of conventional ships and smart ships are the same. But in the case of smart ships, these risks are handled by machines, sensors, software, and communications systems. In such a scenario, physical safety is assured, by increasing the reliability of on board components.

To achieve a well-functioning of smart ships, highly automated vessels must be both generally and operationally reliable. The components of smart ships such as propulsion machines, auxiliary machines, generators, separators, pumps, and cooling systems are complex and, they demand regular preventative maintenance. In the case of reduced or no manning on board machinery is crucial to achieving an acceptable level of reliability. In such a context, reliability measures such as fault detection and recovery, use of redundant components, and fault avoidance improve the overall reliability.

4.3. Smart Ship Functionality

This section adopts a practical approach to review the functionalities as provided under a so-called 'smart ship' by major global ship-building companies. The objective is to identify different enabling technologies used to construct smart ships.

4.3.1. *Autonomous Operation and Maintenance*

A smart ship shall support autonomy in all of control, navigation, and guidance. The objective of this autonomy is to provide:

(1) Safe implementation of innovative technologies in the application of autonomous and remotely controlled vessel functions.
(2) Recommend the work process by implementing technological solutions that can challenge existing statutory regulations and classification rules.

Safety being the main design parameter, the goal of smart ships is to ensure the safety equivalent to or better than observed in conventional ships. Autonomy in ships can eliminate approximately between 75 to 96 percentages of maritime accidents that are caused by human errors [13,14]. The major shipbuilding companies are exploring various technologies such as AI for ships to make an autonomous one [15–17].

The Yara Birkeland, an inland electric container ship manufacturing company, will be constructing a fully autonomous ship by 2022. Some companies build fully autonomous ships from scratch, while others upgrade existing ships into semi-autonomous systems.

4.3.2. *Fuel Consumption*

One example of the implementation of this functionality is demonstrated by the Kalmar. The Eco Reach stacker's fuel consumption model that uses machine learning techniques have been developed by the Kalmer. It guarantees and measures the fuel consumption of an Eco Reach stacker using customer inputs and identifies patterns of cargo handling.

The European Stena Line ferry operator is collaborating with Hitachi on digital analysis and AI to smarten-up its fleet. They plan to unify the existing digital infrastructure of Stena Line and formulate a transformation plan that will make the Stena Line the leading cognitive shipping business in the world by 2021. The aim is to employ analytical methods based on AI to reduce fuel consumption and environmental effects.

4.3.3. *Shipping Markets and Bunker prices*

Mitsui O.S.K. Lines Ltd. and its subsidiary MOL Information Systems have developed a system to provide reliable forecasts of the shipping markets and, bunker prices. They used big data related to ocean shipping and data analysis techniques to address various business and, maritime issues.

Major shipping companies which include Maersk, Panalpina, and Flexport have introduced AI to simulate human intelligence, and to develop automated methods to resolve a range of issues surrounding the maritime industry. Also, they are developing techniques for better estimation of arrival time.

4.3.4. *Intelligent Remote Control*

Sea Machines Robotics has developed an autonomous control and remote control system for optimizing commercial vessel services. In their recent ice-class container ships, Sea Machines and Maersk have launched AI-powered situational awareness technology. They use computer vision, LiDAR, and perception software on a live ship to improve and update vessel operations. Sea Machines claim that their technology will allow vessel operators to minimize their costs by 40% and improve the efficiency of vessels by 200%.

4.3.5. *Hazard Detection and Avoidance*

The Shone is one of the shipping companies that has robotically fused data from shipboard sensors such as radar and camera to create a picture of the hazard and navigation of ships. They adopted AI and computer vision techniques for this purpose.

4.3.6. *Support Passenger Services*

Autonomous and remote-controlled shipping promises to reduce costs and improve protection for ferries and cruise shipping firms. Tugboats and small ferries are supposed to be the first commercial vessels to operate autonomously. The technology could allow passenger service to be expanded into thinly traveled early morning hours and in rural areas.

Rolls-Royce, another shipping company, has also pulled an autonomous AI-based passenger ship operation, a state-run ferry with an automated docking, which avoided obstacles on a 1-mile-long route. These AI-based systems allow the ship to comfortably withstand snow and heavy winds during harsh winter weather.

4.4. Autonomy Levels

One of the essential features of a smart-ship is its ability to function autonomously. This means the ship can safely steer itself under specific conditions. However, the helmsman, a person who steers the ship, quickly interferes in case of abnormal conditions or emergencies. The autonomous ships use AI paradigms to implement various tasks. The objective is to replace with an artificial sub-system, which accomplishes similar tasks like a human does. Under this context, the artificial system should have the right knowledge and the ability to reason correctly.

Autonomy is categorized into seven levels starting from 0 to 6. The traditional human-based steering mode, i.e., no automation is the base level 0. The top-level 6 corresponds to a fully autonomous system. The first three levels, i.e., Level 0, Level 1, and Level 2 fall into the category called *helmsman monitors the environment*. The second category, referred to as, *systems able to monitor the steering environments*, includes the next levels ranging from Level 3 to Level 5. These levels are detailed below.

4.4.1. *Level 0: No Autonomy or Manual Steering*

This level provides no vehicle control, and it provides only manual steering controls. This level of autonomy offers automated warnings about dangers and errors. The navigating officer gives the command for the desired course, speed, and the rest of all the tasks like steering controls or setting points for the course are performed manually.

4.4.2. *Level 1: Decision Support or Helmsman Assistance*

Helmsman assistance is involved, where the control of some specific tasks can be jointly performed by the helmsman and an automated system. Some small steering tasks are performed by the automated system without human intervention, but everything else is fully under human control. The speed is measured by on-board sensors, but the ship is controlled by a human operator.

4.4.3. *Level 2: Partial Automation or Remotely Operated*

This level of autonomy is also called onboard or shore-based decision support. It provides 'hands-off' capabilities when the automated system takes full control of the ship in terms of accelerating, braking, and steering.

Steering of the route is performed with a route plan through a sequence of desired positions. The ship can automatically take safety actions but the helmsman needs to stay alert at the wheel. The helmsman must monitor the operations and be ready to respond quickly in case of an automated system fails to respond properly or in cases of emergency.

4.4.4. *Level 3: Remote Monitoring or Conditional Automation*

Level 3 automation still requires a helmsman, but the human operator can perform some *safety-critical functions* to the ship, under certain conditions. These functions are useful for critical situations that require a quick response, like emergency braking. A sensor system collects the data from the vessel and its surroundings to propose navigation decisions. The operation center uses the sensor data and radar pictures to remotely propose the navigation decision. Thus at this level, although the hands are off the wheel, helmsmen are still required behind the wheel.

4.4.5. *Level 4: High Automation or Monitored Autonomy*

In this level, a ship calculates the overall decision on navigation and operation by assessing the situation and its consequences. Consequences and risks are countered as much as possible. The ship has a sensor system to collect data about surroundings. The automated system infers control actions from the collected data. A human operator is contacted in case of uncertainty about the interpretation of the situation. This level provides self-steering capabilities to the ship like an auto-pilot mode in an aircraft. For instance, the helmsman may safely go to sleep or leave the helmsman's seat. Level 4 ships are capable of steering themselves in almost every environment and condition but might be programmed not to steer in un-mapped areas or during severe weather conditions.

4.4.6. *Level 5: Full Autonomy*

This level is also called *steering-wheel optional* level. Level 5 means full automation in all conditions. This is reached when the system has total control over the ship and no human intervention is required. An automated system itself calculates the navigation and operation decisions as well as the consequences and risks involved. The machine learns from previous and current events and thus exhibits intelligence.

Table 4.1. Autonomy Level (AL) and Tasks

Autonomy Level	Dynamic Steering Task	Monitoring of Steering Environment	Request to Intervene
0	Human Helmsman	Human Helmsman	Human Helmsman
1	Human Helmsman and Automated System	Human Helmsman	Human Helmsman
2	Automated System	Human Helmsman	Human Helmsman
3	Automated System	Automated System	Human Helmsman
4	Human Helmsman	Human Helmsman	Automated System
5	Human Helmsman	Human Helmsman	Automated System

AI-enabled ships are achievable because of the advancement in technological components. But due to regulations and legal battles, Level 5 vehicles are probably from deployments.

Table 4.1 summarizes the different autonomy levels where dynamic steering task includes:

(1) Operational aspects such as steering, braking, accelerating, monitoring the ship and seaway.
(2) Tactile aspects such as responding to events, determining when to change the path, turn, etc.

A request to intervene is a notification by the automated driving system to a human helmsman that a person should promptly begin or resume performance of the dynamic tasks.

4.5. AI-based Technologies for Smart Ships

In this section, we review various AI and allied technologies used for smart-ship.

4.5.1. *Cognitive Technologies*

Cognitive computing is a combination of different technologies like Artificial Intelligence, Computer Vision, Contextual Awareness, Machine Learning, Neural Networks, Natural Language Processing, and Sentiment Analysis that solve day-to-day problems in the same way humans do. The benefits of cognitive technology go well beyond traditional AI systems. These technologies aim to mimic human intelligence and knowledge by analyzing several factors [18].

The cognitive models are useful to analyze risks for maritime operations. Several methodological frameworks that employ cognitive modeling have been developed for reliability and safety analyses. Simulation of the operator's cognition has been recognized as an appropriate and feasible approach in the field of human-machine interaction. The main benefit of using a cognitive approach is that it is time-saving and cost-effective [19].

Cognitive modeling for ship navigation has been built for simple course-tracking tasks. It uses a maritime simulator for analyzing navigation sessions. The simulation model includes four sub-models, i.e., the ship motion model, the interaction model, the helmsman model, and the navigator model.

(1) The ship motion model concerns the numerical description of the ship's behavior and controls any motion of the ship based on the present setting of the ship parameters.

(2) The interaction model generates all the interactions between the navigator model and the ship motion model, and it attempts to maintain all the environmental and system state variables.

(3) The helmsman model works only to transmit the helm order generated by the navigator model to the ship motion model via the interaction model.

(4) The navigator model controls the simulation of the navigator's cognitive and behavioral process. It describes the navigator's cognitive process obtained from the task analysis and its behavioral implementation involves analytical expressions.

4.5.2. *Deep Learning Technologies*

Decision-making processes in autonomous ships play an important role in ocean autonomy. Hence, the technologies for smart ships should have adequate machine intelligence. A major section of ship intelligence consists of a framework based on deep learning. Deep learning is an artificial intelligence technique that imitates the human brain's workings in data processing and creates data patterns for decision-making. Deep learning captures helmsman behavior and exhibits the system intelligence that can be used to navigate autonomous vessels.

In general, deep learning-based frameworks see the problem of a self-driving vessel a data classification problem. The Convolutional Neural Network (CNN) is a type of neural network in DL-based frameworks that has the capability of simultaneously making both the recognition and detection of the objects from the image, based on an automatically extracted large number of features [18]. DL-based frameworks possess the ability to solve complex image classification problems and can provide an elegant mechanism to capture helmsman behavior [20].

Deep learning based on CNN is a major step towards autonomy and has a human-like ability of pattern recognition and detection. Essentially, their performance depends only on the amount of training data, but in practice, few key problems, such as dealing with ambiguity and failure of sensors, need to be resolved. Deep learning frameworks supported by the decision support layer are also expected to overcome many challenges in ship control.

4.5.3. *Knowledge Representation*

Humans are best at understanding, thinking, and interpreting knowledge. Humans know things that are knowledge, and they carry out different actions in the real world according to their knowledge. But to implement such tasks, ships require mechanisms for knowledge representation and reasoning.

Knowledge Representation and Reasoning (KRR) is the sub-field of AI, which is concerned with thinking and how thought leads to intelligent behavior. Machines utilize the knowledge to solve complex real-life problems like communicating with human beings in natural language.

In a conventional maritime transportation system, situation awareness is carried out by human crews on board, while for autonomous shipping this situation perception is handled or fully managed by the autonomous shipping system for which capabilities like KRR and common sense reasoning are essential.

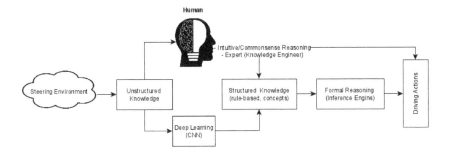

Fig. 4.2.   Knowledge representation in steering actions

The navigator knowledge is defined into two categories- (i) procedural knowledge, referring to the principles of behavior, is mainly contained in all kinds of rules and regulations. (ii) Declarative knowledge obtained by navigators through research, training, and on board ships service is also related to the review and assessment of situations and principles the navigators have developed [21].

As shown in Figure 4.2, the acquired knowledge is recorded and represented to be utilized in a decision support system. The basic tasks of the system include:

(1) Automatic navigation information acquisition and dissemination.
(2) Navigational situation analysis and collision avoidance.
(3) Interaction with the navigator.

Such systems shall perform the following tasks: signaling hazardous situations, current navigational protections based on the guidelines used by experienced navigators, and automatic determination of the device. Further, such systems should perform the following tasks:

(1) Signalling hazardous situations and the current level of navigational safety based on the criteria used by expert navigators.
(2) Control of one or more ship movement operations and trajectories in situations of collision.
(3) The ability to explain and justify the proposed operation.
(4) Present the current navigational situation to the navigator.

Common-sense reasoning is the hidden task that makes AI-based smart ships truly possible. The 90% of the tasks of steering does not need common sense reasoning [22], [23]. To support common-sense reasoning, navigational knowledge from two viewpoints can be considered: competencies and tasks. The first refers to the range of knowledge for management, operational, and support. Second refers to overall transport objective i.e. carriage of cargo and people (voyage planning, loading, passage to the destination, unloading) [24].

4.5.4. *Natural Language Processing*

Natural Language Processing (NLP) is a sub-field of AI that is concerned with the interaction of computers using human language [25], [26]. NLP lets an individual use their normal speech to communicate with machines. NLP is a domain that includes language understanding and manipulation of natural languages. This novel technology can perform much more complex operations, like completing tasks based on past, present, and predicted contexts [27], [28]. Speech recognition is the most preferred method for a human-machine interface in ships after the touch screen interface. It is now being leveraged for a broad spectrum of interactions. Some scenarios are described below-

(1) *Information Updates on News and Weather Reports* Delivery of news is an important task for several reasons on any ship. It informs the crew members about the events happening around them and how they may affect them. News from one country to another is especially important in today's worldwide economy. Knowing what is happening in other countries gives crew members a perspective of the current situation. The current news is particularly important for ships that are going through the water.

Steering a ship is one of the tasks which depends on weather-conditions. Weather plays an important role in the shipping and offshore industry [29], [30]. Traditional ships receive input about weather conditions up to four times a day, via e-mail or HTTP download. Weather maps can be displayed on a screen or printed.

The maritime sector has changed considerably in recent years. Industries are on the edge of implementing the latest technologies for enhancing end-user experiences. Ship manufacturers are increasingly demanding embedded speech solutions in their GPS and navigation systems, as well as their telematics systems. Text-to-speech products enhance the in-ship experience, providing better control over on board systems and maximizing safety, productivity, and overall enjoyment.

Human-like audio from written text improves the customer experience and engagement by interacting with users in multiple languages and tones [31].

(2) *Issue Navigation Commands:* Maritime industries are taking advantage of NLP in designing the navigation system of smart-ships. Voice-based navigation systems play an important role in bridging the gap between man and machine. Speech-recognition systems accept the input in the form of voice commands from the operator and process it to perform the desired action. Upon receiving the input, the NLP component then extracts the navigational related information and use it for mapping route [6].

(3) *Issue Control Commands* NLP allows the ship control systems to respond to voice commands and infer what actions to take, without human intervention. These controls are mainly non-safety-related, such as turning on/off or adjusting lighting, temperature, or open/close door.

(4) *Provide Entertainment Activities:* Play music, turn on/off or search radio programs. The strength of NLP is speech recognition, a procedure of converting speech into words, sounds, and ideas. Simple as it may be, streaming music is one of the key features, largely because it does it so well.

(5) *Implement Assistants:* Personal assistant is a system that performs tasks, or services, on behalf of an individual based on a given input. When a user asks a personal assistant to perform a task, the natural language audio signal is converted into digital data that can be analyzed by the software, and the required task or service is performed. Machines around us can now even talk to us because of NLP. The best part is that they interact with us in our language. Machines not only understand human language, but also reply and perform assigned tasks. Intelligent personal assistants begin to play a more important role in our daily life. A virtual assistant is an artificial intelligence-based technology. It is a combination of various technology: recognition of voices, voice analysis, and speech processing [32], [33].

(6) *Provide Personalized Services:* Setting user account management based on a unique voice pattern. Personalization is a task of understanding user behaviors and the most relevant and well-personalized experience that drives engagement. Many maritime industries are on the path of providing personalization. Original Equipment Manufacturers (OEMs) primarily focus on providing intuitive user experiences. MSC Bellissima, one of the most innovative cruise ships, provides answers to your

questions and easy access to a variety of on board cruise services [34]. When the smart ships become more connected, speech recognition technology can also help minimize accidents, as helmsman can easily send voice command without manually manipulating on board computer systems.

4.6. Cost-Benefit Analysis of Adopting AI for Smart Ships

4.6.1. *Benefits of Smart Ships*

Several major players in the industry are eagerly waiting for smart ships to join their fleet. Smart ships have the potential to revolutionize the shipping industry in terms of designing and operation like the smart-phones did in the telecommunication sector [35]. Deployment of smart-ships in their fleet has the following potential advantages.

(1) *Reduction in Hiring Cost of Crew Members:* The shipping industry is always trying to find ways to lower operating costs. Smaller vessels, where crew costs make up a bigger share of total costs can offer the advantage of reducing the expense of salaries of employees by implementing autonomy. For large ships, the other potential cost savings go beyond the reductions in crew costs.

(2) *Efficiencies of Ships without Crew:* Once we can reduce on board humans and manual work, the entire vessel can be redesigned in a more efficient way that can improve efficiency. The traditional ship where we need to consider a large amount of ship space for the crew can be removed. A traditional ship has a deckhouse, i.e., a cabin on the deck of a ship used for navigation, steering, and accommodation of crew. It would no longer be required in an autonomous ship, thus, open up more space for cargo, possibly making loading and unloading easier. It is not the case where the total crew will be replaced, but, a human may be needed in case of emergencies or if operating conditions become abnormal. With the automation technology, the industries are not visualizing to build the same traditional cargo ship minus crew, but they are planning to give a whole new look starting from scratch.

(3) *Reducing Human Error and Risk:* Just like AutoPilot in aircraft, we can also imagine automation in ships, to avoid human error and risk. Automation is a promising technology that lowers cost related to accidents and insurance by reducing human error or mistakes. According to Allianz Global Corporate & Specialty (AGCS), it is estimated that

between 75% to 96% of accidents in the marine sector are caused due to human error. These incidents due to human error are the primary factor of liability loss [36]. It is not always the case that machines would never make mistakes, but the human error, still a major helmsman of incidents. Big data analysis of crew behavior and other close things could help prevent disasters.

(4) *Reducing the Risk of Piracy:* Piracy endanger the lives of human crew, strongly hinder maritime activity, as well as economic development [37]. Kidnapping crew members for ransom is the main motive behind modern piracy [38]. As per a report from the State of Maritime Piracy [38], there were more than 18 incidents of kidnapping for ransom off the coast of West Africa and more than 21 incidents in Asia. Piracy also costs in the delay of shipment some merchants choose to change their routes due to piracy, which can mean longer delivery times [39]. Meanwhile, with automation in ships, the issue of piracy along certain trade routes would also be expected to be reduced or majorly eliminated. The risk of human crews threatening or holding hostages will also be reduced.

(5) *Sailing over Sea/River/Port:* Humans are involved in performing the group of task related to sailing and coordination of a ship from its departure to arrival at the destination. Human involvement can be reduced by ship automation, which will open new prospects for better, safer, and cheaper operations. The ship will use sensor data to automatically adjust the most optimal sailing route and speed. Another potential benefit of an automated system is safe sailing. A ship can automatically communicate with another ship on the way and thus prevent collision and increased reliability.

(6) *Passing Locks and Bridges:* Currently, the coordination required between the ship and locks and bridges attained through human involvement and communication. Human involvement can be reduced by ship automation that will open new prospects for the finest interaction, safety, and efficient operations for all parties involved. Ships will request a slot to lock/bridge by communicating in accordance with their expected arrival time and sailing schedule. Based on the input received from the ship, the lock/bridge will communicate back to the ship with the intended plan of action. Considering the request for the same lock/bridge by multiple ships in the vicinity, the lock/bridge decision can be made automatically.

(7) *Docking or Departing:* In and out movement of a ship from harbor and docking of the ship is a manual task in coordination with port authorities via interactions. The larger the vessel, the more nautical services are needed for safe dockage/departure, such as piloting, mooring, towing, and coordination. Evolving technologies have the potential to take care of all operations related to docking/departure. Digital coordination systems and magnetic mooring can be used to make the mooring process autonomous and thus optimally control the ship. Smart ships with autonomy can also contribute to reducing fuel consumption by ship and by other nautical services.

(8) *Loading and Unloading:* In a traditional ship, loading, and unloading goods on or from the ship is a manual task. Either port authorities or crew monitored the task. Autonomy in this task can increase productivity and speed, which is roughly equivalent to an expert human performing this task. A stowing and terminal planning system is used for making this task automated. An automated system will communicate with the terminal and ship and perform the task.

4.6.2. *Costs of Deployment of Smart Ships*

Both large and small companies are currently investing money and time in different smart ship related projects. These companies have been facing many challenges. These challenges and the costs associated with them are-

(1) *Inertia to Adopt Smart Ships:* Many of the existing contracts prohibit smart shipping from being adopted because financial rewards encourage full steam ahead rather than dynamics.

(2) *Information Exchange:* The interaction of the ship with the environment is a key element of autonomous shipping. This involves its immediate atmosphere and complex infrastructure and accessing important information such as storage, (amount of room available for stowing materials aboard a ship), while rotating in the port. Too few data are exchanged at the moment. Furthermore, existing data standards are not sufficient to enable the widespread exchange of data.

(3) *Regulations:* Current regulations can be a limiting factor for inland shipping in particular. For example, there are stringent rules about the number of people that need to be on board all the time. Less specific regulations and more general guidelines are required. As technology

advances, regulators will need to adjust the laws to keep pace with the changes in the industry.

(4) *Ship Design:* Ships must be less conventional and more autonomous in architecture. For example, autonomous ships may require special engines and screws, the introduction of electronics and intelligence components, and interchangeable parts, components, which can be upgraded.

(5) *The Human Factor:* In spite of complete autonomy, human beings will continue to play an important role in the functions of the ship, particularly in the event of an emergency. While, in general, the role of the operator may be based on monitoring activities and focuses on software, ensuring that humans are still a vital part of the equation would be important.

4.7. Conclusion

The objective of this chapter is to describe the role of AI and allied technologies in building smart ships. The process of constructing a smart ship begins with identifying a set of design parameters. The chapter presents an initial list parameter that needs to be considered during the design process and a set of functionalities that need to be implemented in smart-ships.

It identifies the reliability of functionalities supported by the ship and safety of passengers, goods, as well as ship itself are the most important design concerns. Any technology that is enabling the construction of the smart-ship idea needs to be evaluated against these parameters.

From this perspective, the chapter presents a cost-benefit analysis of AI and allied technologies. The effectiveness of AI and allied technologies in implementing many functional tasks has been recognized by all the stakeholders in the shipping industries. However, concerns regarding the human aspect and legal barriers need to be overcome.

References

1. M. Levinson, *The Box: How the Shipping Container Made the World Smaller and the World Economy Bigger.* Princeton University Press, 2016.
2. M. Schiaretti, L. Chen, and R. R. Negenborn, "Survey on autonomous surface vessels: Part i-a new detailed definition of autonomy levels," in *International Conference on Computational Logistics.* Springer, 2017, pp. 219–233.

3. L. E. van Cappelle, L. Chen, and R. R. Negenborn, "Survey on short-term technology developments and readiness levels for autonomous shipping," in *International Conference on Computational Logistics*. Springer, 2018, pp. 106–123.

4. P. Martimo and P. D. A. Paalumäki, "Disruptive innovation and maritime sector," Master's thesis, University of Turku, 2017.

5. A. Lazarowska, "Research on algorithms for autonomous navigation of ships," *WMU Journal of Maritime Affairs*, vol. 18, no. 2, pp. 341–358, 2019.

6. P. Withanage, T. Liyanage, N. Deeyakaduwe, E. Dias, and S. Thelijjagoda, "Voice-based road navigation system using natural language processing (nlp)," in *2018 IEEE International Conference on Information and Automation for Sustainability (ICIAfS)*. IEEE, 2018, pp. 1–6.

7. M. Latarche, Ed., *Ship Command and Control*. Adam Foster, 2019.

8. S. Hammoud, "Ship motion control using multi-controller structure," *Journal of Maritime Research*, vol. 9, no. 1, pp. 45–52, 2012.

9. R. Allan, "The Ramora Tugboat Design," Also available as https://ral.ca/d esigns/tugboats/, 2016, [Online; accessed February-2020].

10. S. Adams, "ReVolt next generation short sea shipping," Also available https: //www.dnvgl.com/technology-innovation/revolt/index.html, 2014, [Online; accessed February-2020].

11. K. Maritime, "Autonomous Shipping," 2014, [Online; accessed February-2020 from https://www.kongsberg.com].

12. Kongsberg, "Automated Ships Ltd and Kongsberg to Build First Unmanned and Fully Autonomous Ship for Offshore Operations," 2016, [Online; accessed February-2020 from https://www.kongsberg.com/].

13. O. Levander, "Autonomous ships on the high seas," *IEEE spectrum*, vol. 54, no. 2, pp. 26–31, 2017.

14. T. Schröder, "Autonomous ships - Fact or fiction?" *If's Risk Management Journal*, 2017.

15. Z. Xiao, X. Fu, L. Zhang, W. Zhang, M. Agarwal, and R. S. M. Goh, "Marinemas: A multi-agent framework to aid design, modelling, and evaluation of autonomous shipping systems," *Journal of International Maritime Safety, Environmental Affairs, and Shipping*, vol. 2, no. 2, pp. 43–57, 2019.

16. G. Koikas, M. Papoutsidakis, and N. Nikitakos, "New technology trends in the design of autonomous ships," *International Journal of Computer Applications*, vol. 178, pp. 4–7, 06 2019.

17. A. L. S. Bjørn Johan Vartdal, Rolf Skjong, "Remote-controlled and autonomous ships in the maritime industry," DNVGL-Maritime, Tech. Rep., 2018.

18. S. Ionita, "Autonomous vehicles: from paradigms to technology," in *IOP Conference Series: Materials Science and Engineering*, vol. 252. IOP Publishing, 2017, p. 012098.

19. K. Itoh, T. Yamaguchi, J. P. Hansen, and F. Nielsen, "Risk analysis of ship navigation by use of cognitive simulation," *Cognition, Technology & Work*, vol. 3, no. 1, pp. 4–21, 2001.

20. L. P. Perera, "Autonomous ship navigation under deep learning and the challenges in colregs," in *ASME 2018 37th International Conference on Ocean, Offshore and Arctic Engineering*. American Society of Mechanical Engineers Digital Collection, 2018, pp. 1–10.
21. Z. Pietrzykowski and J. Uriasz, "Knowledge representation in a ship's navigational decision support system," *Marine Navigation and Safety of Sea Transportation*, pp. 71–76, 2009.
22. A. B. Markman, *Knowledge representation*. Psychology Press, 2013.
23. E. Davis, *Knowledge representation*. International Encyclopedia of the Social & Behavioral Sciences: Second Edition, 2015.
24. C. Scrapper and S. Balakirsky, "Knowledge representation for on-road driving," in *Proceedings of the 2004 AAAI Spring Symposium on Knowledge Representation and Ontologies for Autonomous Systems*, 2004, pp. 1–7.
25. K. Lazarevich, "The Future of Smart Devices with Natural Language Processing," Also available as https://www.experfy.com/blog/the-future-of-smart-devices-with-natural-language-processing, 2018, [Online; accessed December-2019].
26. D. Delmolino, "How NLP drives government innovation," Also available as https://www.accenture.com/_acnmedia/Accenture/Redesign-Assets/DotCom/Documents/Local/1/Accenture-Natural-Lan guage-Processing-Digital.pdf#zoom=50, Accenture Federal Services, Tech. Rep., 2019.
27. S. Das, A. Dutta, T. Lindheimer, M. Jalayer, and Z. Elgart, "Youtube as a source of information in understanding autonomous vehicle consumers: natural language processing study," *Transportation research record*, vol. 2673, no. 8, pp. 242–253, 2019.
28. D. Symonds, "Natural language processing enhances autonomous vehicles experience," *https://www.autonomousvehicleinternational.com/*, 2019, [Online; accessed January-2020].
29. M. Della and K. Ajith, "Optimum ship weather routing using gis," Ph.D. dissertation, Kerala University of Fisheries and Ocean Studies, 09 2015.
30. V. Gershanik, "Weather routing optimisation–challenges and rewards," *Journal of Marine Engineering & Technology*, vol. 10, no. 3, pp. 29–40, 2011.
31. A. Trilla, "Natural language processing techniques in text-to-speech synthesis and automatic speech recognition," *Departament de Tecnologies Media*, pp. 1–5, 2009.
32. P. Imrie and P. Bednar, "Virtual personal assistant," in *ItAIS 2013*. AIS Electronic Library (AISeL), 2013, pp. 1–9.
33. P. Imrie, "Virtual personal assistants-a different approach to supporting the end user." in *STPIS@ CAiSE*, 2017, pp. 106–109.
34. K. Holland, "Introducing Zoe, the first personal on-board cruise assistant powered by artificial intelligence," *https://www.telegraph.co.uk/*, 2019, [Online; accessed January-2020].

35. E. Jokioinen, "Remote and autonomous ships the next steps," Advanced Autonomous Waterborne Applications Partners, Tech. Rep., 2016, also available as https://www.rolls-royce.com/~/media/Files/R/Rolls-Royce/documents/customers/marine/ship-intel/aawa-whitepaper-210616.pdf.

36. D. S. Gerhard, "Safety and Shipping," Allianz, Tech. Rep., 2012, also available as https://www.agcs.allianz.com/content/dam/onemarketing/agcs/agcs/reports/AGCS-Safety-Shipping-Review-2012.pdf.

37. L. Quốc Tiến and C. Nguyen, "Impact of piracy on maritime transport and technical solutions for prevention," *International Journal of Civil Engineering and Technology*, vol. 10, pp. 958–969, 01 2019.

38. Giles Noakes and Carly Meredith, Jens Vestergaard Madsen, "The State of Maritime Piracy 2015 - Assessing the Human Cost," Also available as https://oneearthfuture.org/sites/default/files/State_of_Maritime_Piracy_2015.pdf, 2015, [Online; accessed December-2019].

39. M. Schoeman and B. Haefele, "The relationship between piracy and kidnapping for ransom," *Insight on Africa*, vol. 5, pp. 117–128, 07 2013.

Chapter 5

Agent Based Simulation Model for Energy Saving in Large Passenger and Cruise Ships

Eirini Barri[1] , Christos Bouras[1] , Apostolos Gkamas[2] ,
Nikos Karacapilidis[3] , Dimitris Karadimas[4] , Georgios Kournetas[3] ,
Yiannis Panaretou[4]

[1]*Department of Computer Engineering and Informatics,
University of Patras, Patras, 26504 Rio, Greece
E-mail: ebarri@ceid.upatras.gr, bouras@cti.gr, (Christos Bouras is the
corresponding author)*

[2] *University Ecclesiastical Academy of Vella, Ioannina, 45001, Greece
E-mail: gkamas@aeavellas.gr*

[3]*Industrial Management and Information Systems Lab, MEAD,
University of Patras, Patras, 26504 Rio, Greece
E-mail: karacap@upatras.gr, kgiorgos837@gmail.com*

[4]*OptionsNet S.A., 121 Maizonos Str., Patras, 26222, Greece
E-mail: karadimas@optionsnet.gr, panaretou@optionsnet.gr*

5.1. Introduction

Undoubtedly, energy saving is of paramount importance in the shipping industry, as far as both the protection of the environment and the reduction of the associated operating costs are concerned. In this direction, the International Maritime Organization (IMO) aims to reduce ship emissions by at least 50% by 2050, while ships to be built by 2025 are expected to be a massive 30% more energy efficient than those built some years ago [1].

A particular ship category is that of large passenger and cruise ships, which reportedly consume a large amount of energy and thus constitute an interesting area for investigating diverse energy consumption and energy saving solutions. It is estimated that a large ship burns at least 150 tons of fuel per day and emits more sulfur than several million cars, more NO_2 gas than all the traffic passing through a medium-sized town and more particulate emissions than thousands of buses in London [2]. While the cruise ship industry starts taking its first steps toward an emission-free cruise, cruise travels are among the most carbon-intensive ships in the tourism industry; the contribution of the cruise industry to global CO_2 emissions was estimated to be 19.3 Mtons annually in 2010 [3].

To the best of our knowledge, while energy-saving solutions have been thoroughly investigated in the case of (smart) buildings, very limited research [4] [5] has been conducted so far for the above-mentioned ship category. Aiming to contribute to this research gap, this chapter reports on the development of a novel approach that builds on a sophisticated agent-based simulation model for the management of diverse energy consumption issues in large passenger and cruise ships. The model takes into account the size, characteristics (e.g., age, special needs, etc.), and behavior of the different categories of passengers on-board, as well as the energy-consuming facilities and devices of a ship. The application is generic enough to cover requirements imposed by (i) different types of vessels, by taking into account detailed spatial data about the layout of the decks of a ship and the associated position of the energy-consuming devices and facilities; (ii) alternative ship operation modes, corresponding to cases such as the ship cruising during day or night, or being stopped at a port; (iii) different passenger groups in terms of their size and behavior, by considering that the energy consumption of many devices or facilities (e.g., restaurant, air conditioner, etc.) depends on the number of passengers in them or nearby. The proposed agent-based simulation model has been implemented with the use of the AnyLogic simulation software (https://www.anylogic.com/), which provides a nice graphical interface for modeling complex environments and allows the extension of its simulation models through Java code.

A novelty of our approach concerns the exploitation of the outputs obtained from multiple simulation runs by prominent Machine Learning (ML) algorithms to extract meaningful patterns between the composition of passengers and the corresponding energy demands in a ship. In this way, our approach is able to predict alternative energy consumption scenarios and trigger insights concerning the overall energy management in a ship. In addition, it handles the underlying uncertainty and offers highly informative visualizations of energy consumption.

The work reported in this chapter is carried out in the context of the ECLiPSe project (http://www.eclipse-project.upatras.gr), which aims at leveraging existing technological solutions to develop an integrated energy consumption and energy-saving management system for the needs of large passenger and cruise ships. A major task of the project concerns the development of efficient algorithms for the analysis and synthesis of the associated multi-faceted data, which may considerably enhance the quality of the related decision-making issues during the operation of a vessel. These algorithms may trigger recommendations about the management of energy consumption, and accordingly, enable stakeholders to gain energy saving insights.

The remainder of this chapter is organized as follows: Section 5.2 presents the related work aiming to justify and highlight the particularities of our approach. Section 5.3 describes the proposed approach that builds on the synergy of simulation and ML. Section 5.4 explains the data model and the selection of all the constants and variables applied, while Section 5.5 presents in detail the proposed system architecture. Section 5.6 presents indicative experiments and corresponding results from the application of the proposed approach and the analysis of the associated data through appropriate ML algorithms. Section 5.7 comments on the associated data analysis and synthesis issues. Finally, Section 7.8 discusses concluding remarks and Section 5.9 briefly reports on future work directions.

5.2. Background

As mentioned in the previous section, while considerable research has been conducted so far on the optimization of various energy consumption issues in buildings (whether they are smart or not), very limited work has been reported so far in the case of large ships. Due to this reason, the work discussed in this section concerns related approaches in (smart) buildings or parts of them (e.g., offices). As a general remark, we note that many of these approaches are based on simulation models and start utilizing ML algorithms.

A representative case of an agent-based model for office energy consumption is described in [6]. This work elaborates the elements that are responsible for energy consumption and presents a mathematical model to explain the energy consumption inside an office. The proposed model is validated through three sets of experiments giving promising results.

Energy consumption and emission of the maritime industry is studied in [7]. The authors introduce an artificial neural network model in order to explore the sailing data, aiming to predict the fuel consumption for cruise ships. Authors aim to minimize the fuel consumption, achieving the economic and environmental protection of a voyage using optimization algorithms. It is demonstrated that their method and tool can be used to plan the sailing speed of cruise ships in advance.

Adopting another perspective, a review of ML models for energy consumption and performance in buildings is presented in [8]; the motivation of this work was the exploitation of contemporary technologies, including network communication, smart devices, and sensors, toward enhancing the accuracy of prediction in the above energy management issues. In a similar research direction, a combination of mathematical statistics and neural network algorithms to solve diverse energy consumption problems is proposed in [9]; this work analyzes the associated big data aiming to facilitate energy consumption predictions for various types of buildings.

Using a Gaussian process, the researchers in [10] developed a model to predict the fuel consumption under different circumstances, running the model in different scenarios. In the introduced model, the effects of speed and trim and the impact of the wind and waves were considered and evaluated. The study indicated the accuracy and also the efficiency of using the Gaussian process for energy consumption prediction.

Another study based on energy consumption and gas emissions is described in [11]. A comparison has been made between inland river shipping and seagoing ships. The authors took data in calm water and real navigation conditions and calculated at the end the energy efficiency operation. The results show that the environment can affect significantly the operational energy efficiency of ships.

A comparative analysis of energy-saving solutions in buildings appears in [12]; the proposed tool for assessing the effectiveness of energy-saving technologies implementation allows not only to evaluate individual decisions, but also to compare and rank them according to the breakeven rate for the efficient implementation decline. A combination of Nearest Neighbors and Markov Chain algorithms is described in [13] for the implementation of a system that is able to support decision making about whether to turn on or off a device in a smart home setting, thus handling the related energy management issues.

As argued, one should thoroughly analyze the nature of available or collectible data and the particular application, to choose the most suitable approach. In any case, ML algorithms may enable stakeholders to gain insights from energy usage data obtained under different scenarios. For instance, the ML-based smart controller for a commercial building's HVAC (heating, ventilation, and air conditioning) system that is described in [14] managed to reduce its energy consumption by up to 19.8%. In [15], authors use different methods of artificial intelligence to estimate the fuel a ship consumes during a trip. The authors have taken data from a commercial ship and divided them into training and data set. Using multiple linear regression, the computer has been taught to estimate data that are not known. The predictions made by ML are finally compared against real data.

The author in [16] aims at filling a gap in the existing scientific knowledge on the way energy in its different forms is generated, converted, and used on-board a vessel. This is done by applying energy and exergy analysis to ship energy system analysis. The results of this analysis allow improving the understanding of energy flows on-board and identifying the main inefficiencies and waste flows.

A different perspective is adopted in [17] which elaborates the reduction of energy consumption in pumping stations. By comparing the energy consumption of a station during a 15-days-period and what would the station consume in the same period after the energy audit, one may quantify the gain resulting from the use of the proposed management system. After examining the existing state of a pumping station, evaluating its energy performance, and developing improvement actions, this work proposes a set of solutions to improving energy consumption.

The authors in [18] present state of ship energy consumption evaluation in China and abroad is analyzed. A mathematical modeling method of fuzzy evaluation is adopted to evaluate ship energy consumption. Six indexes for fuzzy evaluation are set up, and the weight vectors of the index system in level one and level two evaluation are both determined through expert investigations. The energy consumption evaluation of a ship is carried out. It is indicated by analysis that fuzzy evaluation is suitable for evaluating ship energy consumption. The result is instructive for ship energy consumption evaluation.

A new method to model the ship energy flow and thus understand the dynamic energy distribution of the marine energy systems is introduced in [19] using the MATLAB/Simscape environment, a multi-domain simulation method is employed. As reported, the proposed method can help people better monitor the ship's energy flow and give valuable insights about how to efficiently operate a vessel. In a similar research line, aiming to provide a better understanding of the use of energy, the purpose it serves, and the efficiency of its conversion on-board, an analysis of the energy system of a cruise ship operating in the Baltic Sea is provided in [20] based on a combination of direct measurements and computational models of the energy system of the ship, the proposed approach ensures to provide a close representation of the real behavior of the system.

5.3. Overall Approach

The work reported in this chapter is carried out in the context of a two-year research project, which comprises of four major phases, namely (i) analysis of energy consumers in large passenger and cruise ships, (ii) development of methods to analyze and process the associated data, (iii) development of basic services for the visual representation of energy consumption, and (iv) development of an innovative platform to facilitate the related decision-making process. Through these phases, the project will develop contemporary methods to gather, aggregate, and analyze heterogeneous data representing both the energy consumption in diverse devices and facilities and the concentration of passengers in different areas of a ship. In addition, the project will develop a set of novel services aiming to optimize the management of energy consumption. Finally, the project will produce a set of guidelines for energy saving in a ship.

Our approach adopts the action research paradigm [21], which aims to contribute to the practical concerns of people in a problematic situation; it concerns the improvement of practices and strategies in the complex setting under consideration, as well as the acquisition of additional knowledge to improve the way shipping stakeholders address issues and solve problems. Building on the strengths of existing related work, as reported in the previous section, the proposed approach comprises two main phases: (i) agent-based simulation of the energy consumption in various sites of a ship, and (ii) utilization of prominent ML algorithms on the outputs of multiple simulation runs to extract meaningful insights about the relation between the passenger composition and corresponding energy demands. Through these phases, our approach is able to gather, aggregate, and analyze heterogeneous data representing both the energy consumption in diverse devices and facilities and the concentration of passengers in different areas of a ship.

To fine-tune our approach, a series of meetings with shipping companies were conducted;. Through them, we identified the types of devices and facilities that mainly affect energy consumption in the ship categories under consideration and obtained valuable information concerning the parameters to be taken into account in energy consumption models (such as that energy supply in a ship is provided by a number of electric power generators, which in most cases are of different capacity and do not work in parallel). In addition, the information collected concerned the layout of ship decks and its relation to the energy management issues investigated. Finally, we clarified issues related to the alternative types of passengers and how these may influence alternative energy consumption and energy-saving scenarios.

5.3.1. *Agent-Based Simulation*

Our approach aims to enable stakeholders to predict the energy needs of a ship (e.g., to recommend the appropriate number of power generators to operate each time), facilitate predictive maintenance issues (affecting the related equipment), and, hopefully, reduce energy-related operating costs. This approach facilitates the modeling of energy consumption, especially for ships that do not have sophisticated energy consumption monitoring and control systems. To fulfill these aims, our agent-based simulation model takes into account the passengers' behavior and its dependencies with a ship's facilities, devices, and resources.

A basic assumption of our approach is that the energy demands in many sites of a ship (such as the restaurant, nightclub, the kids' daycare, etc.) depend on the number of passengers who gather at these sites at a given time, as well as their composition in terms of type (customer or crew member), age, gender, etc. We consider that different age groups have different paths and habits (differences among passenger groups may even affect the speed of a moving agent). To estimate the populations gathered in these sites, we relied on the behavioral preferences of large sub-groups of passengers. For instance, we assume that young passengers prefer to spend their time at a nightclub from 10 pm to 3 am, while elderly passengers prefer to eat dinner at a fancy restaurant. Our model may also simulate the behavior of Persons With Special Needs (PWSN); in particular, we assume that these people move at a slower pace and are in most cases accompanied by another person. Such assumptions enable us to predict the gathered populations and, accordingly, the energy demands during day and night. This approach facilitates the modeling of energy consumption, especially for ships that do not have sophisticated energy consumption monitoring and control systems.

In addition, according to our approach, the passengers' behavior is considered and modeled through three basic scenarios corresponding to the ship (i) being moved during the day, (ii) being moved during the night, and (iii) being anchored at a destination or port. In the above scenarios, we assume different behaviors from passengers, which may result in different energy demands. Finally, to accommodate the spatial particularities of each ship, our approach pays much attention to the layout of each deck. These layouts provide us with the spatial data that is needed to calculate the movement of passengers inside the ship. AnyLogic offers a user-friendly import of sectional plans (views), thus enabling the production of a more realistic model of the distribution of ship passengers, facilities, and devices. Taking into account what our models predict in terms of energy needs, we suggest different policies of energy management, aiming to reduce energy consumption.

5.3.2. *Machine Learning Algorithms*

Machine Learning (ML) is an application of Artificial Intelligence (AI) that enables information systems to automatically learn and improve from past data, without being explicitly programmed. ML includes diverse approaches such as supervised, unsupervised, and reinforcement learning.

Having thoroughly assessed the palette of broadly used ML algorithms for the needs of our approach, we decided to utilize two classification algorithms, namely, the Decision Trees (DT) and the K-Nearest Neighbors (K-NN) algorithms. This is due to the fact that these algorithms provide high interpretability of their results,have low computational cost, and fit well into our data structure.

DT is one of the simplest and widely used classifiers in the field of data mining. It constitutes a non parametric supervised learning method, aiming to create a model that predicts the value of a target variable by learning simple decision rules inferred from the data features. DT demonstrates excellent applicability in data sets with either categorical or continuous variables. In addition, it requires little data preparation and it is able to process large amounts of data [22].

K-NN is a simple supervised ML algorithm that can be used for both classification and regression problems and has been extensively applied in diverse disciplines, such as economics and health [23]. It relies on labeled input data to learn a function that produces an appropriate output when given new unlabeled data. In most cases, K-NN yields competitive results and has significant advantages over other data mining methods. It differs from other classifiers in that it does not build a generic classification model; instead, whenever a new record is being inserted into the system, it tries to find similar records (nearest neighbors) from past data stored in its memory and assigns it the value of the dependent variable that its neighbors have.

5.4. Data Model

One of the main variables to be taken into account concerns the passengers on board, as well as their composition in terms of type (customer or crew member), age, gender, etc. As mentioned above, different age groups have different paths and habits; for example, elderly passengers go to sleep earlier than the kids, so the corridors have to have enough light to satisfy both groups. The difference between the age groups affects also the speed that an agent can have, which will be also different from a person with special needs. Another important variable considered is the ship layout (at the deck level). These layouts provide information about the detailed coordinates of all ship locations, including passenger cabins and facilities of the ship. Each energy-consuming device is also a variable in our model.

5.5. System Architecture

Through user-friendly interfaces, our approach enables stakeholders to build and run alternative energy consumption scenarios. These scenarios are populated with data that are either given by the user or already stored in the application's repository. The execution of scenarios is through the AnyLogic simulation engine, which results in the creation of illustrative reports and associated energy-saving directions ('recipes'). A middleware component establishes the connection between the application's back-end and front-end, while also enabling the interoperability of the proposed application with external services.

Figure 5.1 sketches the components and overall architecture of the proposed application. As shown, the four main components (microservices) of the proposed application are:

- The service that describes the scenario and the specifications of the desired energy modeling and analysis. Initially, this service specifies all the necessary specifications of the AnyLogic simulation tool models. In a future version, this service will include data from smart energy consumption meters or actual passenger position/movement.

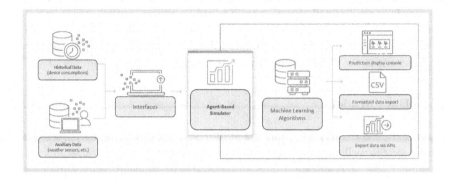

Fig. 5.1.   Architecture of the proposed application

- The basic simulation service of energy consumption scenarios ,which service's is responsible for the simulation of various energy consumption scenarios in order to support managers of the passenger ship draw conclusions about energy consumption optimization.
- The results display service, which provides simulation results in various formats including raw data reports, comparison charts, etc.
- The data analysis service, which relies on ML algorithms and tries to detect trends and behaviors regarding energy consumption and possible ways to optimize it.

5.6. Experiments

This section illustrates a particular set of experiments carried out to assess the applicability and potential of our application for a specific vessel. In particular, we elaborate on energy demands that are associated with four popular facilities of a ship, namely, the (i) night-club, (ii) kids'daycare, (iii) casino, and (iv) restaurant. For the case under consideration, we consider and import in the simulation software the original deck layouts, where all ship facilities and passenger cabins are mapped. Moreover, we assume a total population of 3100 passengers on-board, belonging to four distinct age groups (i.e., 1-14, 15-34, 35-54, $\geq$55 years old). Table 5.1 summarizes sample data concerning the populations of each age group in the facilities considered. For each group of passengers, we create a simple linear behavioral model in which each group remains in a specific facility for some time. We do this for every group of passengers and every period to create a comprehensive routine for all passengers throughout the day. In this way,

Table 5.1. Distribution of age groups in various ship facilities

Ship's Facility	Age Group	1-14	15-34	35-54	≥55
Nightclub	Percentage	0	60	30	10
	Population	0	300	150	50
Kids'daycare	Percentage	35	10	55	0
	Population	53	15	82	0
Restaurant	Percentage	12	8	35	45
	Population	46	30	134	172
Casino	Percentage	0	0	35	65
	Population	0	0	112	208

we can simulate diverse scenarios, which may be easily aggregated to create an illustrative energy consumption map for the whole vessel.

5.6.1. *Nightclub*

For the case elaborated in this chapter, we generated random samples of 500 passengers, assuming that the percentage of passengers visiting this facility is between 15 and 17%. This facility operates from 11 pm to 5 am. The conditional probability of someone visiting the night club is shown in Table 5.1. We also set the time spent there (from passengers of all age groups) to follow a triangular distribution with a lower limit equal to 50 minutes, mode equal to 95 minutes, and upper limit equal to 110 minutes. Finally, we imported the layout of a specific deck, where detailed spatial data about the cabins and the possible pathways leading to the night club area are described. By running the corresponding simulations, we are able to visualize the possible concentration of passengers during the night at this area of the ship (see Figure 5.2).

Consequently, by estimating the energy requirements of the night club with respect to the number of passengers hosted, we can calculate the possible energy needs for the particular time period and facility (see Figure 5.3). Such estimations can be used for future predictions of energy consumption in cases where passengers are distributed in a similar way. Furthermore, the derived data can be statistically analyzed to reveal the data patterns and mechanisms that may cause the particular energy demands.

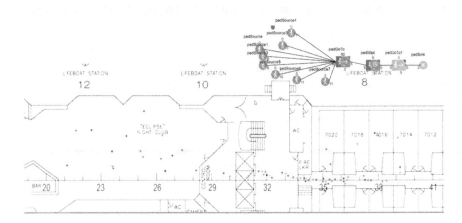

Fig. 5.2.   Energy demands corresponding to passenger concentration in the nightclub

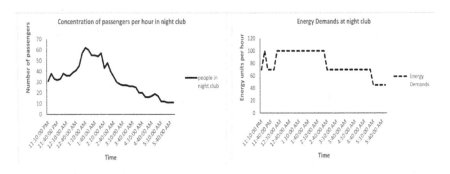

Fig. 5.3.   An instance of a simulated energy consumption scenario in the nightclub

5.6.2. *Kids'Daycare*

For this facility (see Figure 5.4), we considered that the passengers who visit it are mainly children (1-14 years old) and their parents (who may belong into the age groups of 15-34 and 35-54 years old). The opening hours of this facility are from 11 am to 2 pm. We assumed that the daycare is not the only choice that the above groups have for entertainment purposes. Also, compared to other areas on the ship, the daycare is not large enough to accommodate all parents with their children. We have therefore assumed that the proportion of passengers visiting it daily ranges from 4 to 5.5%, i.e., from 120 up to 176 persons. The time people spend while visiting

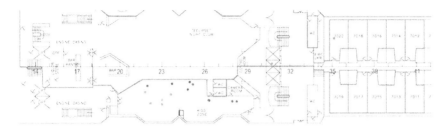

Fig. 5.4. An instance of a simulated energy consumption scenario in the kids'daycare

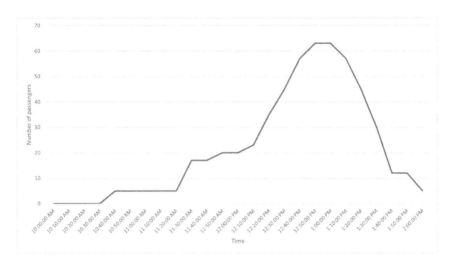

Fig. 5.5. Passenger concentration in the kids'daycare

this facility is described by a triangular distribution with a minimum of 50 minutes, maximum of 110 minutes, and dominant value of 80 minutes. The experiments carried out gave the concentration of passengers as shown in Figure 5.5.

5.6.3. *Casino*

The samples of passengers used in the particular set of experiments concerned 320 people (i.e., 10% of average passenger population). We assumed that this facility operates from 7 pm to 7 am and mainly attracts passengers that are older than 35 years old (65% of them belonging to the ≤55 age group and the remaining 35% to the 35-54 age group). Moreover, passengers that visit the casino are divided into two categories: those who

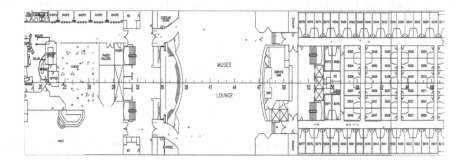

Fig. 5.6.   An instance of a simulated energy consumption scenario in the casino

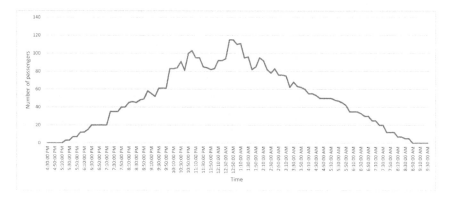

Fig. 5.7.   Passenger concentration in the casino

choose to spend their time exclusively in the casino during the night (20%) and those who visit the casino for a certain time period (they may leave and re-enter the casino during the night). The first category concerns 20% of the casino visitors (their stay follows a triangular distribution with a minimum of 250 minutes, maximum of 300 minutes, and dominant value of 270 minutes). Similarly, for the rest 80% of casino visitors, we considered that their time spent follows a triangular distribution with a minimum of 20 minutes, maximum of 80 minutes, and dominant value of 35 minutes).

As illustrated in Figure 5.7, there is a two-peak distribution of passenger concentration. This kind of distribution is called bimodal distribution because of the existence of two distinct modes. In our approach, the "bimodal distribution" is the after-effect of our assumption that passengers visiting the casino can be classified into two different sub-groups.

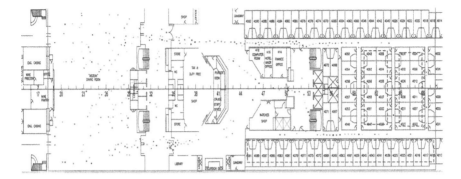

Fig. 5.8.   An instance of a simulated energy consumption scenario in the restaurant

5.6.4. *Restaurant*

We considered one of the available ship restaurants (offering an la carte menu, thus not being an economic one), operating from 7 pm to 11 pm. This facility concerns all passengers, regardless of age group. We assumed that 10-12% of passengers (320-380 people) choose this particular restaurant; their stay is described by a triangular distribution with a minimum of 75 minutes, maximum of 150 minutes, and dominant value of 120 minutes.

Figure 5.8 depicts an over-concentration of passengers at a main ship's corridor leading to the particular restaurant. This phenomenon can be identified as a problem of bad operation scheduling in one of the ship's cites that could possibly cause inconveniences and/or delays in passengers' service. Such problems can be diagnosed by running the appropriate simulation models under the right assumptions.

Figure 5.8 depicts passenger concentration in the restaurant area.

5.7. Data Analysis and Synthesis

The experiments described above demonstrate diverse features and options offered by the proposed simulation model. To predict energy consumption in large passenger and cruise ships, our approach aggregates results obtained from each particular facility of a ship and produces a corresponding time series diagram, in which the dependent variable is the energy consumption measured in energy units per hour and the time interval is 10 minutes. Figure 5.10 illustrates the overall energy demands with regards to the estimated gathering of passengers in the facilities discussed in the previous section throughout the day. Obviously, our experiments have not

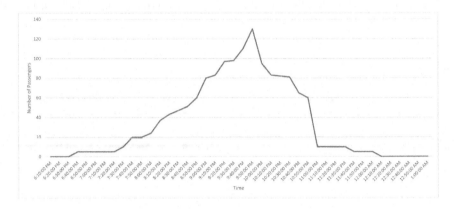

Fig. 5.9.   Passenger concentration in the restaurant

Table 5.2.   Time slots considered
in our approach

Time Interval	Time Slot
7:00am 11:59am	Morning
12:00pm 4:59pm	Midday
5:00pm 9:59pm	Evening
10:00pm 6:59am	Night

considered the entirety of facilities and energy consumers available on a ship (such as air conditioning, lighting, heating, etc.); however, all of them can be easily aggregated to our model and thus provide a detailed mapping of the overall energy consumption.

Building on the proposed agent-based simulation model that facilitates the creation of alternative energy consumption scenarios, we can produce realistic data that can be further elaborated by prominent ML algorithms to provide meaningful insights for managing diverse energy consumption patterns [24]. Parameters taken into account by the proposed ML algorithms also include the number of ship generators (categorical variable), alternative age groups and their populations (as defined for each ship), and time slots considered each time (the ones adopted in our approach are shown in Table 5.2).

In our experiments, we generated a large data set of 919 different passenger compositions for each time slot. A small sample of this data set,

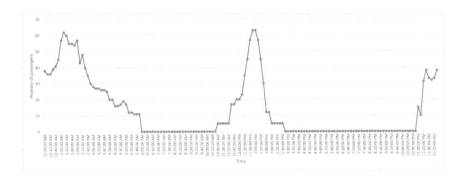

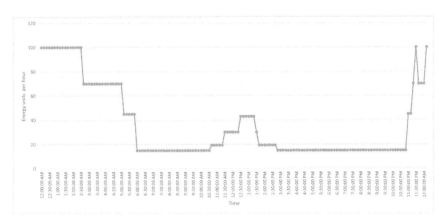

Fig. 5.10. Cumulative concentration of passengers in four major ship facilities and corresponding energy demand

concerning only four of these compositions for the time slots defined, is presented in Table 5.3 (the number of generators that operate for each data combination is calculated upon the definition of a set of energy unit intervals and their association with the energy produced by the simultaneous operation of a certain number of generators). A big part of this data set (70%) was used as the training set of the two ML algorithms incorporated in our approach. Through the utilization of these algorithms, one may predict the required number of generators per time slot for a specific passenger composition.

Focusing on the 'morning' time slot, Figure 5.11 illustrates the output of the Decision Tree algorithm, which classifies alternative passenger compositions into different numbers of power generators required. As it can be observed, the energy consumption of the ship in this time slot is affected by

Table 5.3. Sample of our data set.

Composition ID	Age Group				PWSN*	Time Slot	Number of Generators in Simultaneous Operation
	1-14	15-34	35-54	≥55			
1	290	535	945	1432	97	Morn.	3
						Mid.	2
						Even.	4
						Night	3
2	200	750	1200	1100	75	Morn.	2
						Mid.	3
						Even.	4
						Night	4
3	175	700	1150	1150	20	Morn.	2
						Mid.	3
						Even.	4
						Night	4
4	48	885	1890	550	100	Morn.	1
						Mid.	3
						Even.	4
						Night	3

*PWSN : people with special needs.

(i.e., positively correlated to) the ratio of passengers that are older than 55 to those that are younger than 35 years old. The interpretation of this may be that older people are more active in the morning (compared to younger populations). Results shown in Figure 5.12 provide additional evidence in favor of the above insight; as depicted, the correlation between the number of generators used in the morning and the number of elderly passengers is positive.

For the above-mentioned time slot, we also applied the K-NN algorithm. The confusion matrix produced (this matrix is actually a technique for summarizing the performance of a classification algorithm) showed us limited reliability. In particular, K-NN performed very well with more than 95% accuracy when classifying compositions of passengers that were associated with the operation of one or four generators, while this was not the case for compositions associated with the operation of two or three generators; in these cases, the accuracy was about 45 and 55%, respectively.

Table 5.4 summarizes a small set of predictions produced by the K-NN algorithm for the cases of one of four generators operating simultaneously.

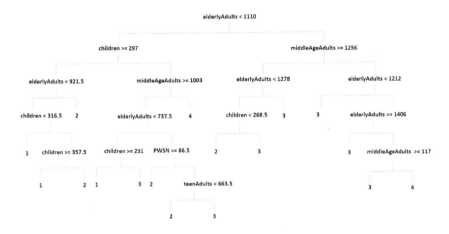

Fig. 5.11. Decision Tree classification ('children', 'teenAdults', 'middleAgeAdults' and 'elderlyAdults' correspond to the 1-14, 15-34, 35-54 and ≥55 age groups, respectively)

It is noted that for these cases K-NN produces very similar results to those obtained by the Decision Tree, i.e., the energy needs are positively correlated to the ratio of passengers who are older than 55 to those who are younger than 35 years old.

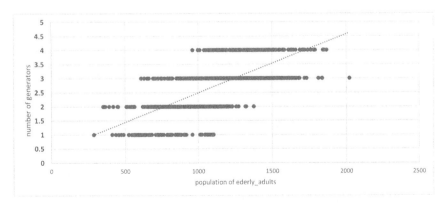

Fig. 5.12. Scatter plot of the number of generators vs. the population of elderly passengers

Table 5.4. Predictions produced by K-NN algorithm

Age Group				PWSN	Number of Generators in Simultaneous Operation
1-14	15-34	35-54	≥55		
100	755	1100	1300	75	4
270	668	916	1570	43	4
174	968	865	1021	40	4
243	755	1412	656	41	1
328	686	1450	678	82	1
410	995	1425	780	10	1

5.8. Conclusion

The application described in this chapter advances the way stakeholders of large passenger and cruise ships deal with energy consumption issues, by building on a comprehensive and informative simulation model that facilitates the creation and assessment of alternative energy saving scenarios. We argue that our overall approach suits particularly ships that are not equipped with state-of-the-art (smart) energy management sensors and devices. To accommodate this situation, our approach produces realistic data that can be analyzed and give insights for the mechanisms of energy consumption.

The prediction of energy consumption in large passenger and cruise ships is certainly a hard problem. This is mainly due to the need to simultaneously consider the interaction between multiple parameters and agent behaviors. To deal with this problem, the proposed approach blends the process-centric character of a simulation model and the data-centric character of ML algorithms.

By building on a comprehensive and informative agent-based simulation model, our approach facilitates the generation and assessment of alternative energy consumption scenarios that incorporate vast amounts of realistic data under various conditions. Moreover, it advocates the use of prominent ML algorithms to aid the finding, understanding, and interpretation of patterns that are implicit in this data, ultimately aiming to provide meaningful insights for shaping energy-saving solutions in a ship.

5.9. Future Work

The proposed approach in this chapter promotes the way cruise ships deal with energy consumption. It provides the management of the cruise ship with a lot of information that can facilitate the reduction of energy waste.

One step toward this solution should be the production of realistic data that can be analyzed to trigger insights about the mechanisms of energy consumption. The projected energy requirements should formulate a set of management rules for dealing with energy consumption at the energy-saving stage. In addition, the production of realistic data can lead us to produce more sophisticated results by applying ML algorithms and finally being able to promote the ideal energy management policy regarding passenger composition.

Our future work includes the comparison of the outputs of the proposed approach with real data. As far as the outcomes produced by the agent-based simulation model are consistent with real data, our ML algorithms will be better trained, which will in turn enhance the accuracy of the associated energy consumption predictions. Such reinforcement learning activities consist of one of our future work directions. Other directions include the investigation of alternative modes to combine simulation and ML. Specifically, we plan to consider the application of ML algorithms prior to and within the simulation. In the former case, we will need real data to develop rules and heuristics that our agent-based simulation model can then employ. In the latter, we may reuse previously trained ML-based models or train the ML models as the simulation is taking place.

References

1. Current awareness bulletin. International Maritime Organization. Accessed 2020-01-23. [Online]. Available: http://www.imo.org/en/KnowledgeCentre/CurrentAwarenessBulletin/Documents/CAB\%20258\%20MAY\%202018.pdf
2. Large cruise ship and supersized pollution problem. The Guardian. Accessed 2020-01-23. [Online]. Available: https://www.theguardian.com/environment/2016/may/21/the-worlds-largest-cruise-ship-and-its-supersized-pollution-problem
3. E. Eijgelaar, C. Thaper, and P. Peeters, "Antarctic cruise tourism: the paradoxes of ambassadorship,'last chance tourism' and greenhouse gas emissions," *Journal of Sustainable Tourism*, vol. 18, no. 3, pp. 337–354, 2010.
4. E. Barri, C. Bouras, A. Gkamas, N. Karacapilidis, D. Karadimas, G. Kournetas, and Y. Panaretou, "Towards an informative simulation-based application for energy saving in large passenger and cruise ships," in *2020 6th IEEE International Energy Conference (ENERGYCon)*. Gammarth, Tunisia: IEEE, Oct. 2020, pp. 117–121.
5. ——, "A novel approach to energy management in large passenger and cruise ships: Integrating simulation and machine learning models," in *10th International Conference on Simulation and Modeling Methodologies,*

Technologies and Applications. [Online]. Available: http://telematics
.upatras.gr/telematics/publications/novel-approach-energy-management-l
arge-passenger-and-cruise-ships-integrating-simulation-and-machine

6. T. Zhang, P.-O. Siebers, and U. Aickelin, "Modelling office energy consump-
    tion: An agent based approach," *SSRN Electronic Journal*, 2010, available
    at: http://dx.doi.org/10.2139/ssrn.2829316.

7. J. Zheng, H. Zhang, L. Yin, Y. Liang, B. Wang, Z. Li, X. Song, and
    Y. Zhang, "A voyage with minimal fuel consumption for cruise ships," *Jour-
    nal of Cleaner Production*, vol. 215, pp. 144–153, 2019.

8. S. Seyedzadeh, F. P. Rahimian, I. Glesk, and M. Roper, "Machine learning
    for estimation of building energy consumption and performance: a review,"
    Visualization in Engineering, vol. 6, no. 1, pp. 1–20, 2018.

9. S. Guzhov and A. Krolin, "Use of big data technologies for the implementa-
    tion of energy-saving measures and renewable energy sources in buildings,"
    in *2018 Renewable Energies, Power Systems & Green Inclusive Economy
    (REPS-GIE)*. Casablanca, Morocco: IEEE, Apr. 2018, pp. 1–5.

10. J. Yuan and V. Nian, "Ship energy consumption prediction with gaussian
    process metamodel," *Energy Procedia*, vol. 152, pp. 655–660, 2018.

11. X. Sun, X. Yan, B. Wu, and X. Song, "Analysis of the operational energy
    efficiency for inland river ships," *Transportation Research Part D: Transport
    and Environment*, vol. 22, pp. 34–39, 2013.

12. Y. Chebotarova, A. Perekrest, and V. Ogar, "Comparative analysis of effi-
    ciency energy saving solutions implemented in the buildings," in *2019 IEEE
    International Conference on Modern Electrical and Energy Systems (MEES)*.
    Kremenchuk, Ukraine: IEEE, Sep. 2019, pp. 434–437.

13. R. G. Rajasekaran, S. Manikandaraj, and R. Kamaleshwar, "Implementa-
    tion of machine learning algorithm for predicting user behavior and smart
    energy management," in *2017 International Conference on Data Manage-
    ment, Analytics and Innovation (ICDMAI)*. Pune, India: IEEE, Feb. 2017,
    pp. 24–30.

14. A. Javed, H. Larijani, and A. Wixted, "Improving energy consumption of a
    commercial building with iot and machine learning," *IT Professional*, vol. 20,
    no. 5, pp. 30–38, 2018.

15. T. Uyanik, Y. Arslanoglu, and O. Kalenderli, "Ship fuel consumption predic-
    tion with machine learning," in *Proceedings of the 4th International Mediter-
    ranean Science and Engineering Congress*, Antalya, Turkey, Apr. 2019, pp.
    25–27.

16. F. Baldi, "Improving ship energy efficiency through a systems perspective,"
    Ph.D. dissertation, Chalmers University of Technology, Gothenburg, Sweden,
    2013. [Online]. Available: http://publications.lib.chalmers.se/records/fullt
    ext/186189/186189.pdf

17. M. Kaddari, M. El Mouden, A. Hajjaji, and A. Semlali, "Reducing energy
    consumption by energy management and energy audits in the pumping sta-
    tions," in *2018 Renewable Energies, Power Systems & Green Inclusive Econ-
    omy (REPS-GIE)*. Casablanca, Morocco: IEEE, Apr. 2018, pp. 1–6.

18. Y. Guohao, X. Yiqun, and L. Rongmo, "Fuzzy evaluation of ship energy consumption," *Navigation of China*, vol. 4, pp. 22–25, 2011.

19. G. Zou, M. Elg, A. Kinnunen, P. Kovanen, K. Tammi, and K. Tervo, "Modeling ship energy flow with multi-domain simulation," in *Proc. 27th CIMAC World Congress on Combustion Engines. International Council on Combustion Engines*, Shanghai, China, May 2013.

20. F. Baldi, F. Ahlgren, T.-V. Nguyen, M. Thern, and K. Andersson, "Energy and exergy analysis of a cruise ship," *Energies*, vol. 11, no. 10, p. 2508, 2018.

21. P. Checkland and S. Holwell, "Action research: its nature and validity," *Systemic Practice and Action Research*, vol. 11, no. 1, pp. 9–21, 1998.

22. O. Z. Maimon and L. Rokach, *Data mining with decision trees: theory and applications*. World Scientific, 2008.

23. P. Cunningham and S. J. Delany, "k-nearest neighbour classifiers," *Multiple Classifier Systems*, vol. 34, no. 8, pp. 1–17, 2007.

24. T. M. Deist, A. Patti, Z. Wang, D. Krane, T. Sorenson, and D. Craft, "Simulation-assisted machine learning," *Bioinformatics*, vol. 35, no. 20, pp. 4072–4080, 2019.

PART 3

Safety and Security in Smart Ships

Chapter 6

Infectious Disease and Indoor Air Quality Management in a Cruise Ship Environment

Bernard Fong[1,*]

[1] *College of Computing and Informatics, Providence University, Taiwan Blvd Sec. 7 No. 200, Taichung 433, Taiwan (e-mail: bfong@ieee.org).*
* Bernard Fong is the corresponding author.

A. C. M. Fong[2]

[2] *Department of Computer Science, Western Michigan University, 1903 W Michigan Ave., Kalamazoo MI 49008, USA (e-mail: afong@wmu.edu).*

C. K. Li[3]

[3] *Alpha Positive Clinic 23/F, New World Tower 1, 18 Queen's Road, Central, Hong Kong (e-mail: enckli@iconnect.polyu.hk).*

6.1. Introduction

The recent global COVID-19 outbreak has exposed the vulnerability of indoor environmental control on-board cruise ships. Modern giant cruise ships can host as many as 9,000 passengers and crew members in total. With thousands of peoples on-board, air monitoring becomes an extremely important topic to cover, as ventilation is reported to be among the most effective in infectious disease control [1]. Disease prevention, as well as general wellness, is becoming an increasing concern when hundreds of patients are infected in one single ship. While the size of cruise ships has grown to increase guest capacity, the health risks associated with accommodating thousands of people while sailing on open waters also increase as even one asymptomatic pathogen carrier can infect many [2]. On-board a cruise ship, environmental factors have a substantial impact on crew and passenger wellness and this Chapter takes a look at how Internet of Things (IoT) and smart technologies can efficiently control air quality and minimizes the risk of spreading communicable diseases.

In recent years, smart technologies have been widely used in many cruise ships central to an on-board WiFi network. The popularity of smart phones and sensing technology can offer a diverse range of solutions for a plethora of services. The use of individual smart phones effectively forms a network of a distributed system through combining computational powers with sensing networks that can alleviate many of the challenges associated with interoperability in having a single system for both entertainment and control [3]. In addition, wireless connectivity such as WiFi, ZigBee, and Bluetooth Low Energy (BLE) can provide efficient communication links between passenger smart phones and on-board sensors [4].

The design of mega ships that host thousands of people simply cannot cope with any substantial increase in social separation between guest and crew member, so any technologies that attempt to solve challenges associated with infectious disease and indoor air quality (IAQ) management need to reduce the likelihood of a large number of people gathering in the same area within the ship. Control of air quality and the importance of minimizing the risk of infectious disease spread has been an important topic to study for a long time [5]. Optimizing IAQ across all areas of a cruise ship, therefore, becomes a vitally important task to ensure the safety of all passengers and crew on board. To achieve this goal, this chapter takes an in-depth look into how smart technology can be deployed to tackle these challenges.

6.2. Crowd Control and Communicable Disease in Cruise Ships

As observed in many pandemics over recent years, the vast volume of human traffic between international borders has posed immense challenges to controlling infectious diseases. This is particularly challenging for cruise ship operators, as many guests board from different countries. It is vitally important to enable appropriate resource management to prevent pandemic outbreaks when the ship sails out to open waters. As demonstrated by the complete shutdown of many cruise lines in early 2020, current disease surveillance systems lack the ability to detect and deny embarkation to guests that are asymptomatic pathogen carriers.

Many occasions can draw hundreds or even thousands of people to gather in an indoor environment such as while awaiting disembarkation that happens almost every morning. This can substantially increase the risk of spreading communicable diseases [6]. Cutting the waiting time and eliminating queues are certain measures for minimizing the risk of infection. A WiFi-centered solution can easily help put an end to most of the queues on-board a cruise ship. A number of venues where guests usually queue up can make use of WiFi-based notifications so that they can wait inside their stateroom instead of lining up together.

On-board IoT system connects a diverse range of smart devices, generically known as a thing that is distributed across the entire ship. In addition to crowd control, combining big data analytics and Artificial Intelligence (AI) also allows cruise ship operators to gain insights into the passengers' spending behavior [7]. IoT not only handles crowd control, but also serves as sales promotion. For example, discount coupons can be offered to selected passengers in or near the shop's area.

The basic implementation of the smart ship IoT system is shown in Fig. 6.1. While this system is fairly simple and serves multiple purposes in crowd control both for the reduction in risk of spreading infectious diseases and increase in sales of merchandise, there are a number of deployment considerations that need to be thoroughly addressed. These include data heterogeneity from various data sources, lack of ubiquitous connectivity and data security as well as interoperability and cross-platform compatibility.

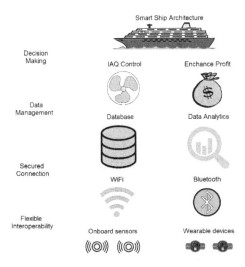

Fig. 6.1. Basic ship IoT system

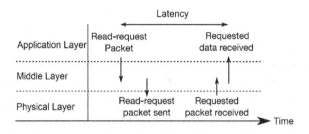

Fig. 6.2. Request-response process timeline.

Smart services and applications are often delivered through different protocols, such as IEEE 802.11 Wi-Fi, BLE, and ZigBee, which possess major challenges for integrating different services on-board. The time needed to poll data in a request-response cycle, defined as the latency, is perhaps the most important factor in service integration. Latency measurement on the application layer is illustrated in Fig. 6.2.

6.3. Automation for Indoor Air Quality Management

As a significant number of cruise ships have been put on suspension during the first quarter of 2021, there is an urgent need of developing an automated framework that provides effective mitigation strategies and environmental management in an indoor environment. Syndromic surveillance approaches for detection by monitoring disease symptoms are particularly in need where crowds are gathered in confined closed spaces during certain events and activities [8].

6.3.1. *Infectious Disease Spread Risks*

There are several health issues pertaining to infection risk that can be significantly elevated by people gathering. This includes inadequate ventilation, separation distance from a pathogen carrier, and the unpredictable movement of people while participating in activities such as dancing and games. Figure 6.3 is only one of numerous examples where a substantial number of guests gather in a confined space of the ship. In this kind of situation where guests participate in activities where maintaining a social distance adequate for preventing pathogen spread through someone sneezing among the crowd, careful control of ventilation to filter out any aerosols in the air becomes the only practical solution to minimize the risk of disease spread.

Fig. 6.3. Crowd with minimal social distance gathering inside a confined venue on a cruise ship

6.3.2. *Controlling IAQ for Minimizing the Risk of Spreading Infectious Diseases*

Traditionally, disease outbreak forecasting relies on applying standard monitoring charts such as, Exponentially Weighted Moving Average (EWMA), Cumulative Sum (CUSUM), or Shewhart, which are commonly used in various industrial applications [9]. In the ship environment where a substantial number of people can gather in confined indoor spaces, the challenge in minimizing the risk of spreading infectious disease would be acquisition and analysis of disease incident data that would be multiple data types by nature [10]. Dynamically adjusting ventilation to optimize IAQ would entail interrogating disparate data sets with such complexity and diverse conditioned data sets.

Simulation studies are a significant part of pathogen-spread scenario prediction, thereby enabling enhancing accuracy for disease spread modeling, which is vital for mitigation and containment of infectious diseases through effective airflow control. Some disease-spread simulation models aim to understand the effects of changes in people's behavior or movement [11]. Symptom detection enables the prediction of disease emergence and spread patterns based on symptoms and behavioral model analysis. One such example would be the emergence and subsequent spread of the influenza virus [12]. As the disease emerges at an early stage, an abnormally high number of patients with flu-like symptoms used in addition to spatial and temporal information about confirmed cases can be well-used as precursors for early detection of a major outbreak.

Detection and early warning of symptoms inlcude coughing and sneezing through real-time data collection with pattern recognition, health monitoring, and management through prognostics analysis [13]. The most common existing methods for outbreak detection include temporal and spatial surveillance [14]. Syndromic surveillance can be supported through an on-board IoT system as described next.

6.4. IoT in a Cruise Ship

IoT devices have been widely used in monitoring a wide range of environmental data such as air temperature and pressure, relative humidity, luminosity as well as noise level [15]. The collected data can be sent to a server via WiFi for processing and environmental control. Recent research has reported effective usage of IoT platform for ventilation control in buildings [16] and school campuses [17], yet very little has been done on optimizing IoT solutions for ships despite the extensive use of connected smart technologies on-board many modern ships.

We investigate the use of IoT and smart technologies to dynamically control the IAQ of public areas. First, a prognostics approach similar to that widely used in system diagnostics can be adopted for crowd estimation [18]. Ventilation is controlled based on the estimation of crowd size as well as historical data of people traffic from past events.

Broadly speaking, a smart ship IoT architecture, extending from the basic system shown earlier in Fig. 6.1, can be viewed as a 5-layer architecture as shown in Fig. 6.4. The top layer, Layer-5, is an IoT sub-system with two parts of users, namely, staff with varying functions and passengers that can either be of equal status or ranked according to stateroom or membership grades. This layer identifies the user class within the entire IoT system, i.e., physically anywhere across the ship. The primary function of this top layer is user class identification for services [19]. The next layer manages all collected data at Layer 4 from acquisition to processing. One of the major functions here is event management which, in the case of IAQ control, concerns ventilation for different events [20]. Layer 4 deals with data associated with the ambient environment, power consumption as well as people concentration and location. All the data here is closely linked to infectious disease spread prevention and IAQ management. The middle layer deals with wireless connectivity at Layer 3, where different types of devices from on-board to those brought onto the ship by crew members and guests are connected to the ship's network. One important design consideration

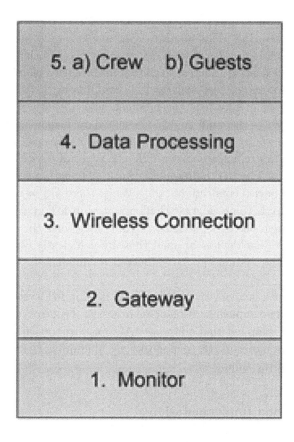

Fig. 6.4.   Smart ship IoT architecture

is scalability [21] and security [22]. Due to the diverse nature of devices, the heterogeneous nature of ship IoT implies that wireless connectivity can involve WiFi, Bluetooth, ZigBee, etc. on-board, as well as external links such as satellite or cellular like 4G and 5G when the ship is sufficiently close to land.

Layer 2 is the gateway that serves as a link between the users and their devices. The gateway may perform tasks such as protocol conversion and edge service applications [23]. It is responsible for uploading data as well as network resources allocation from different devices to the ship's servers. Also, information retrieval for end-users, such as data access and Internet surfing is supported by this layer.

At the bottom is Layer 1 that links different devices together for the purpose of monitoring. This monitor layer serves a wide range of devices which include mobile and wearable devices that people bring on-board, as well as fixed devices of different facilities like security cameras, ambient sensors, smoke detectors, lighting, and air conditioning [24].

In the context of ship IoT, the inherent heterogeneity of devices, networks, and services demands highly efficient data management. Further, device interoperability and cross-platform compatibility must be considered to facilitate data exchange between different types of devices and systems across the ship. The IoT backbone network, therefore, needs to support heterogeneous device connectivity [25]. When there are so many types of devices, security becomes a great challenge to ship IoT system implementation, the system becomes particularly vulnerable when the ship is docked as it is possible that devices on land that are sufficiently close can access the system. The system can be subjected to different types of security threats [26].

The backbone network is crucial within the ship IoT system, as it supports the dynamic implementation of environmental control such as lighting and air quality. Ship IoT that addresses IAQ optimization can be built upon a network of air monitors; these monitors are ubiquitous facilities across all indoor areas of the ship.

6.5. Design and Implementation

6.5.1. *Ship IoT Framework*

The system architecture of ship IoT where each air monitor within the ship is considered as a connected smart node is shown in Fig. 6.5. Here, all the air monitor nodes are connected autonomously within a mesh network to a single gateway. Wireless code dissemination is an important design consideration for ship IoT configuration management since the IoT devices can be updated periodically [27], thereby making the network self-cognizant by using a multicast-based Over-The-Air (OTA) approach. Given that there are so many air monitors throughout the ship and their low-bandwidth requirements and a large number of devices, make such design option particularly suited for dynamic monitoring [28]. The gateway distributes a series of multicast OTA packets to all the air monitors across the network.

The IAQ control function serves as part of the overall ship IoT system, where the overall system also connects to a large variety of smart devices,

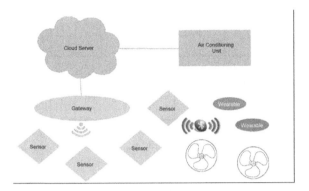

Fig. 6.5.   Ship air quality monitoring IoT module.

such as lighting, smoke alarms, security monitors, energy consumption monitors, etc., that combined form a smart ship system in a modular form. The infectious disease and IAQ management module can be considered as part of the overall ship IoT network. The module utilizes air monitors in a meshed wireless network to gain connectivity throughout the ship and the main purpose is to optimize IAQ through adjusting air ventilation.

6.5.2. *Self-Cognizant Prognostics Air Control*

Wireless code dissemination is elementary for IoT configuration management across the ship such that each IoT device can be self-cognizant in situ [29]. Self-cognizant air control utilizes multicast-based OTA prognostics approach given the properties of low-bandwidth and the number of IoT devices within the module shown in Fig. 6.5. Here, the shipping gateway receives an OTA command from the cloud upon detecting an anomaly [30], for example, when a gas sensor detects CO^2 level exceeds the pre-set threshold or the number of wearable devices increases, it indicates that more people are within a certain area. As this happens, the gateway multicasts a series of OTA packets to all the nodes that are related to air control. The airflow can be increased or decreased according to the environment.

6.5.3. *Case Study: Controlling Hydraulic Pump Fan System*

We take a look at how self-cognizant prognostics is implemented in the control of a hydraulic pump fan system. In this case study, the ventilation system covers one section of 16 staterooms, as illustrated in Fig. 6.6, where eight staterooms are inside cabins that air circulation is virtually controlled

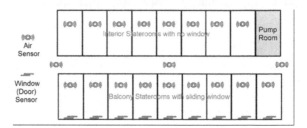

Fig. 6.6.   Section of a stateroom deck where IAQ is controlled through a self-cognizant prognostics approach

by the fan system; whereas the other side of the corridor consists of nine balcony staterooms, where ventilation also takes into account the opening and closing of a sliding glass door that is not a simple binary opening or closing but also senses how wide an individual door is opened.

When controlling an individual fan, the fan blade rotates at a varying speed that regulates the airflow according to the IAQ condition. In the hydraulic pump section, there is a piston that reciprocates in its cylinder bores, which controls the intake as well as the discharge of air [31]. The inclined angle of the swashplate controls the displacement of the pump, which is controlled by stroking piston discharge pressure [32]. The airflow is in turn controlled by adjusting the swashplate angle [33]. Moving the piston in the axial direction is controlled by concentrating pressurized air through the piston neck toward the hydrostatic balance area. While controlling the pump, the inlet pressure is also controlled to minimize the cavitation that can cause flow instability [34]. The control of discharge flow and air pressure through the piston pump is shown in Fig. 6.7. When the detected IAQ is at an acceptable level, i.e., does not exceed the preset threshold, the rate of airflow is roughly maintained around 16 to 17 liters per minute while the discharge pressure is below the preset pressure of approximately 2 KPa, and the airflow stops when the discharge pressure exceeds the preset pressure. This is to automatically save energy when the system detects the absence of any person in the area.

One of the major applications of self-cognizant prognostics is automatic fault diagnosis and self-healing [35]. During the time when many cruise ships were severely affected by the global pandemic in early 2020, it has been shown that minimizing maintenance work has a very positive impact on the safety of crew members.

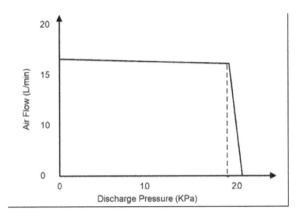

Fig. 6.7. Regulation of airflow with no anomaly detected.

Looking further into the relationship between IAQ control and system operational reliability, and system performance degradation due to issues such as component wear and abrasion through analysis of Physics of Failure (PoF) would result in a reduction in inlet pressure [36]. Stability of airflow depends on the control of discharge pressure [37], as illustrated in Fig. 6.8 that shows the simulation result when inlet pressure drops abnormally due to system failure. During normal operation, both the swashplate and the transmission shaft rotate coaxially on the same central axis. By using a self-prognostics algorithm, swashplate can adjust itself when an offset in the radial direction is detected through automatic centering so that it can

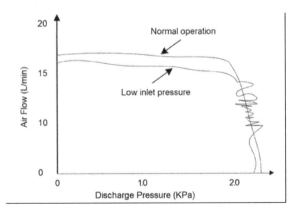

Fig. 6.8. Impact on airflow control due to insufficient inlet pressure

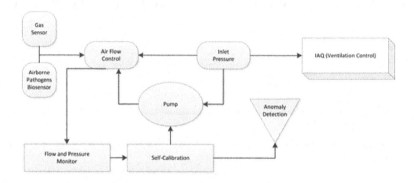

Fig. 6.9. Smart self-diagnosis system

return to the central axis, this would ensure constant airflow without any manual intervention in the event of minor misalignment.

6.5.4. *Smart Self-Diagnosis System*

In addition to connecting user devices to the ship, the on-board IoT system also facilitates the collection of sensor signals for the purpose of monitoring system performance and to carry out self-diagnosis for various components including those associated with IAQ management. A smart self-diagnosis system consists of sensors, data collection in conjunction with signal conditioning sub-systems. The self-diagnosis system shown in Fig. 6.9 consists of an airflow rate sensor, a pair of vibration sensors, as well as a pressure sensor. The airflow rate sensor is used to measure the output of the fan in liters per minute. Each of the vibration sensors monitors the vibration of the pump in one direction, namely axial and radial, and the pressure sensor monitors the pump's discharge. The signal conditioning sub-system that consists of filter circuitry is a vital part for running self-diagnosis algorithms and also to supply power to all four sensors. Both the airflow sensor and pressure sensor are directly connected wirelessly to the ship's on-board IoT system. Airflow will automatically increase when more people enter the section. Clustering algorithm analyzes people's movement by controlling the sum of the relative speed of individual fans within the section [38]. This airflow computation process continuously controls ventilation as people move around within the section concerned. When no one is in the section, the system saves energy by lowering the inlet pump pressure of the pump to zero so that all fans stop. As people enter the section, airflow increases gradually.

6.6. Conclusion

This chapter looked at how IoT can be used in managing the air quality dynamically to minimize the risk of infectious disease spreading in a ship environment where thousands of people move around. In addition to providing IAQ management across the ship, ship IoT is becoming an integral part of the smart ship by supporting a wide variety of services. We looked at the use of syndromic surveillance and prognostic algorithms for optimizing IAQ through smart control of ventilation in indoor areas so that the risk of pathogen spread within the ship can be minimized.

References

1. A. D. Hill, R. A. Fowler, K. E. Burns, L. Rose, R. L. Pinto, and D. C. Scales, "Long-term outcomes and health care utilization after prolonged mechanical ventilation," *Annals of the American Thoracic Society*, vol. 14, no. 3, pp. 355–362, 2017.
2. G. Suleyman, G. Alangaden, and A. C. Bardossy, "The role of environmental contamination in the transmission of nosocomial pathogens and healthcare-associated infections," *Current Infectious Disease Reports*, vol. 20, no. 6, pp. 1–11, 2018.
3. C. Dameff, B. Clay, and C. A. Longhurst, "Personal health records: more promising in the smartphone era?" *JAMA*, vol. 321, no. 4, pp. 339–340, 2019.
4. S. Mumtaz, A. Alsohaily, Z. Pang, A. Rayes, K. F. Tsang, and J. Rodriguez, "Massive internet of things for industrial applications: Addressing wireless iiot connectivity challenges and ecosystem fragmentation," *IEEE Industrial Electronics Magazine*, vol. 11, no. 1, pp. 28–33, 2017.
5. J. M. Villafruela, F. Castro, J. F. San José, and J. Saint-Martin, "Comparison of air change efficiency, contaminant removal effectiveness and infection risk as iaq indices in isolation rooms," *Energy and Buildings*, vol. 57, pp. 210–219, 2013.
6. N. Zhang, W. Chen, P.-T. Chan, H.-L. Yen, J. W.-T. Tang, and Y. Li, "Close contact behavior in indoor environment and transmission of respiratory infection," *Indoor Air*, vol. 30, no. 4, pp. 645–661, 2020.
7. K. P. Subbu and A. V. Vasilakos, "Big data for context aware computing–perspectives and challenges," *Big Data Research*, vol. 10, pp. 33–43, 2017.
8. A. T. Fleischauer and J. Gaines, "Enhancing surveillance for mass gatherings: the role of syndromic surveillance," *Public Health Reports*, vol. 132, no. 1 suppl., pp. 95S–98S, 2017.
9. R. J. Carnevale, T. R. Talbot, W. Schaffner, K. C. Bloch, T. L. Daniels, and R. A. Miller, "Evaluating the utility of syndromic surveillance algorithms for screening to detect potentially clonal hospital infection outbreaks," *Journal of the American Medical Informatics Association*, vol. 18, no. 4, pp. 466–472, 2011.

10. C. J. E. Metcalf, K. S. Walter, A. Wesolowski, C. O. Buckee, E. Shevliakova, A. J. Tatem, W. R. Boos, D. M. Weinberger, and V. E. Pitzer, "Identifying climate drivers of infectious disease dynamics: recent advances and challenges ahead," *Proceedings of the Royal Society B: Biological Sciences*, vol. 284, no. 1860, p. 20170901, 2017.

11. J. A. Tracey, S. N. Bevins, S. VandeWoude, and K. R. Crooks, "An agent-based movement model to assess the impact of landscape fragmentation on disease transmission," *Ecosphere*, vol. 5, no. 9, pp. 1–24, 2014.

12. F. Santermans, K. Van Kerckhove, A. Azmon, W. J. Edmunds, P. Beutels, C. Faes, and N. Hens, "Structural differences in mixing behavior informing the role of asymptomatic infection and testing symptom heritability," *Mathematical Biosciences*, vol. 285, pp. 43–54, 2017.

13. L. Bourouiba, "Turbulent gas clouds and respiratory pathogen emissions: potential implications for reducing transmission of covid-19," *JAMA*, vol. 323, no. 18, pp. 1837–1838, 2020.

14. A. C. Hale, F. Sánchez-Vizcaíno, B. Rowlingson, A. D. Radford, E. Giorgi, S. J. O'Brien, and P. J. Diggle, "A real-time spatio-temporal syndromic surveillance system with application to small companion animals," *Scientific Reports*, vol. 9, no. 1, pp. 1–14, 2019.

15. K. McLeod, P. Spachos, and K. N. Plataniotis, "An experimental framework for wellness assessment using the internet of things," *IEEE Internet Computing*, vol. 24, no. 2, pp. 8–17, 2020.

16. G. Chiesa, S. Cesari, M. Garcia, M. Issa, and S. Li, "Multisensor iot platform for optimising iaq levels in buildings through a smart ventilation system," *Sustainability*, vol. 11, no. 20, p. 5777, 2019.

17. A. A. Hapsari, A. I. Hajamydeen, and M. I. Abdullah, "A review on indoor air quality monitoring using iot at campus environment," *International Journal of Engineering & Technology*, vol. 7, no. 4.22, pp. 55–60, 2018.

18. M. Baybutt, C. Minnella, A. E. Ginart, P. W. Kalgren, and M. J. Roemer, "Improving digital system diagnostics through prognostic and health management (phm) technology," *IEEE Transactions on Instrumentation and Measurement*, vol. 58, no. 2, pp. 255–262, 2008.

19. R. Ivanov, "An approach for developing indoor navigation systems for visually impaired people using building information modeling," *Journal of Ambient Intelligence and Smart Environments*, vol. 9, no. 4, pp. 449–467, 2017.

20. A. Caron, N. Redon, F. Thevenet, B. Hanoune, and P. Coddeville, "Performances and limitations of electronic gas sensors to investigate an indoor air quality event," *Building and Environment*, vol. 107, pp. 19–28, 2016.

21. B. Fong, N. Ansari, A. C. M. Fong, and G. Y. Hong, "On the scalability of fixed broadband wireless access network deployment," *IEEE Communications Magazine*, vol. 42, no. 9, pp. S12–S18, 2004.

22. Y. Xiao, X. Shen, B. Sun, and L. Cai, "Security and privacy in rfid and applications in telemedicine," *IEEE Communications Magazine*, vol. 44, no. 4, pp. 64–72, 2006.

23. N. Cheng, W. Xu, W. Shi, Y. Zhou, N. Lu, H. Zhou, and X. Shen, "Air-ground integrated mobile edge networks: Architecture, challenges, and opportunities," *IEEE Communications Magazine*, vol. 56, no. 8, pp. 26–32, 2018.

24. S. Salimi and A. Hammad, "Critical review and research roadmap of office building energy management based on occupancy monitoring," *Energy and Buildings*, vol. 182, pp. 214–241, 2019.

25. S. K. Roy, S. Misra, and N. S. Raghuwanshi, "Senspnp: Seamless integration of heterogeneous sensors with iot devices," *IEEE Transactions on Consumer Electronics*, vol. 65, no. 2, pp. 205–214, 2019.

26. T. Melamed, *An active man-in-the-middle attack on bluetooth smart devices.* WIT Press, 2018, vol. 8, no. 2, pp. 200–211.

27. S. Ravichandran, R. K. Chandrasekar, A. S. Uluagac, and R. Beyah, "A simple visualization and programming framework for wireless sensor networks: Proviz," *Ad Hoc Networks*, vol. 53, pp. 1–16, 2016.

28. Y. Bejerano, C. Raman, C.-N. Yu, V. Gupta, C. Gutterman, T. Young, H. A. Infante, Y. M. Abdelmalek, and G. Zussman, "Dymo: Dynamic monitoring of large-scale lte-multicast systems," *IEEE/ACM Transactions on Networking*, vol. 27, no. 1, pp. 258–271, 2019.

29. A. Taherkordi, F. Loiret, R. Rouvoy, and F. Eliassen, "Optimizing sensor network reprogramming via in situ reconfigurable components," *ACM Transactions on Sensor Networks (TOSN)*, vol. 9, no. 2, pp. 1–33, 2013.

30. L. Li, K. Ota, and M. Dong, "Deepnfv: A lightweight framework for intelligent edge network functions virtualization," *IEEE Network*, vol. 33, no. 1, pp. 136–141, 2018.

31. T. Hamamoto, S. Omura, N. Ishikawa, and T. Sugiyama, "Development of the electronically controlled hydraulic cooling fan system," *SAE Transactions*, pp. 1439–1446, 1990.

32. B. Xu, Q. Chao, J. Zhang, and Y. Chen, "Effects of the dimensional and geometrical errors on the cylinder block tilt of a high-speed eha pump," *Meccanica*, vol. 52, no. 10, pp. 2449–2469, 2017.

33. M. I. Mahmud and H. M. Cho, "Analysis of forces in an automobile ecv using in external variable displacement swash plate type compressor for air conditioning control system," *Journal of Mechanical Science and Technology*, vol. 28, no. 5, pp. 1979–1984, 2014.

34. J. Tao, H. Wang, H. Liao, and S. Yu, "Mechanical design and numerical simulation of digital-displacement radial piston pump for multi-megawatt wind turbine drivetrain," *Renewable Energy*, vol. 143, pp. 995–1009, 2019.

35. B. Fong, N. Ansari, and A. C. M. Fong, "Prognostics and health management for wireless telemedicine networks," *IEEE Wireless Communications*, vol. 19, no. 5, pp. 83–89, 2012.

36. A. Cubillo, S. Perinpanayagam, and M. Esperon-Miguez, "A review of physics-based models in prognostics: Application to gears and bearings of rotating machinery," *Advances in Mechanical Engineering*, vol. 8, no. 8, pp. 1–21, 2016.

37. S. Rahnama, A. Afshari, N. C. Bergsøe, and S. Sadrizadeh, "Experimental study of the pressure reset control strategy for energy-efficient fan operation: Part 1: Variable air volume ventilation system with dampers," *Energy and Buildings*, vol. 139, pp. 72–77, 2017.

38. X. Ren, C. Zhang, Y. Zhao, G. Boxem, W. Zeiler, and T. Li, "A data mining-based method for revealing occupant behavior patterns in using mechanical ventilation systems of dutch dwellings," *Energy and Buildings*, vol. 193, pp. 99–110, 2019.

Chapter 7

Survey of Security and Privacy of Navigation, Localization, and Path Planning in Smart Ships

Jiacheng Li, Yang Xiao*

Department of Computer Science, The University of Alabama, Tuscaloosa, AL 35487-0290, USA
(e-mails: jiachengli@ieee.org, yangxiao@ieee.org)
* Yang Xiao is the corresponding author.

7.1. Introduction

In ancient marine times, human beings usually utilized logs or air-filled animal skins as small ships to cross small bodies of water. These simple ships can carry large goods easily [1]. With the increasing importance of marine transportation and trade, ships continue to progress. By the end of 1100, more and more large ships improve the handling characteristics of ships through rudders [2]. As years passed, to sail more efficiently and enhance competitively and attractively, smart ships are quietly replacing traditional ships in oceans.

Smart ships can navigate ships to destinations, track ship' locations, or plan paths by utilizing sensors, communication systems, satellite systems, etc. Nowadays, navigation systems have become almost indispensable because they provide enough safety for smart ships. Meanwhile, tracking systems can accurately locate the positions of ships when they send out distress signals or ask for help. Furthermore, according to marine environments and information of ports, smart ships are also able to plan the best route to avoid hitting rocks or collision with other ships. Although every ship uses several aid tools for the utmost safety, these aid tools contain some potential security and privacy issues. We will present these potential issues in this chapter.

In this chapter, we first introduce basic concepts of systems of navigation, localization, and path planning in Section 7.2. Then, we present their security issues of navigation, localization, and path planning, in Sections 7.3, 7.4, and 7.5, respectively. Then, we analyze the privacy concerns for each system in section 7.6. We discuss their countermeasures in Section 7.7. Finally, we conclude this chapter in Section 7.8.

7.2. Concepts of Navigation, Localization, and Path Planning

Systems of navigation, localization, and path planning are essential tools for smart ships. These systems can navigate ships to target positions, monitor the real-time status, provide the shortest and safest routes, and avoid collisions of ships. We will present the concept of navigation, localization, and path planning in detail next.

7.2.1. *Navigation*

A navigation system is an aid instrument that can determine the positions of ships or vehicles and guide users to a particular place. In the process of movement of ships or vehicles, navigation systems can monitor the process of the whole movement and prompt users to change directions at specific positions to reach final destinations. Navigation systems usually include four categories: marine, land, space, and aeronautic [3]. In this chapter, we will introduce marine navigation systems, which are E-navigations. E-navigation is defined as a unified collection, exchange, integration, and analysis of marine information on ships and ashore through electronic means to enhance original positioning to final destination navigation and related services for ensuring maritime safety and security as well as protecting the marine environment.

7.2.2. *Localization*

A localization system is a positioning system that can track, identify, and monitor the real-time positions of ships or vehicles. For smart ships, localization systems can enhance the ability to control and handle ships in the marine environment. For example, a dynamic positioning system can keep the positions of ships in the deep sea by analyzing the wind and the wave data [4]. Meanwhile, due to the influence of natural factors, such as winds, waves, and currents, ships may be forced to deviate from their courses. In this scenario, localization systems can keep ships on course, and will not be taken away by fluctuating winds and waves. On the other hand, localization systems are also beneficial to the management and security of ports. For example, localization systems can monitor the movements of ships to avoid collision and traffic jams in the port.

7.2.3. *Path Planning*

A path planning system is an essential component of automation systems and entails finding the shortest and a collision-free path between a start position and a goal position. Path planning systems, which are an indispensable tool in ships, can help ships avoid collisions with dynamic or static obstacles in any potentially dangerous waters. Meanwhile, through using a path planning system, ships can be autonomously navigated while obeying international maritime laws.

7.3. Security Issues of Navigation in Smart Ships

With modern facilities and automated functions, today's smart ships have several advanced navigation systems, which can provide accurate data for voyages. However, these navigation systems also have potential security issues. We will present their security issues next.

7.3.1. *Global Navigation Satellite Systems*

Global Navigation Satellite System (GNSS) is an umbrella term that describes any satellite navigation system with global coverage. GNSS currently includes the United States' Global Positioning System (GPS), Russia's Glonass, European Union's Galileo, China's Beidou, and Japan's Quasi-Zenith Satellite System. Modern vessels are installed with GNSS modules that provide positioning, navigation, and timing information. The working principle of GNSS is that a set of satellites provide signals to transmit positioning and timing information from space to GNSS receivers [5]. In this scenario, GNSS receivers utilize these data to determine vessel locations. Although GNSS provides a safety guarantee for vessels, crews, and cargoes, jamming and spoofing attacks always threaten the safety of vessels [6]. Jamming and spoofing attacks are described below as follows:

- GNSS Spoofing Attacks: A GNSS spoofing attack involves broadcasting fake GNSS signals or rebroadcasting genuine signals at different times to deceive GNSS recipients [5]. The principle of GNSS spoofing is that attackers broadcast fake GNSS signals that appear the same as genuine signals. When fake GNSS signals are gradually increased, genuine signals can be fully replaced by fake GNSS signals. In this scenario, once targeted vessels are controlled by fake GNSS signals, controlled vessels sail to a position that is determined by attackers. This kind of attack is also called a "carry-off attack".

- GNSS Jamming Attacks: GNSS jamming, which is deliberately disturbing, aims to block or stop GNSS signals through frequency transmitting equipment. Slight GNSS jamming leads to signal loss, while serious GNSS jamming brings prominent risks for public security. For example, since GNSS jammers are not able to distinguish kinds of communications, they may disrupt all connections within a broad frequency range.

7.3.2. *Marine Radar*

A marine radar, which is a mandatory aid that is installed in vessels, is majorly used in vessels' positioning, identifying, tracking, and preventing collisions. To save time from observing a target to obtaining collision avoidance data, a marine radar is usually used with an Automatic Radar Plotting Aid (ARPA). Meanwhile, an ARPA also helps a marine radar to establish tracking by radar contacts. A system of marine radar and ARPA can accurately calculate the tracking object's speed, Closest Point of Approach (CPA), and course. However, weather and environments at sea are always changeable, and these unpredictable situations may cause significant security risks when marine radars are working. These potential risks are as follows:

- *Clutters*: Clutters, which are unwanted echoes that affect marine radar signals, are usually returned from the sea, rain, ground, or atmospheric turbulence, etc., [7]. Since clutters can be generated and returned in any medium, radar operators hardly distinguish whether return signals are generated by clutters or real objects. In the ocean, unpredictable weather is a major reason to cause clutters, which affect marine radar signals. For example, when vessels encountered rainfall or hail at sea, marine radar signals may be attenuated or lost.

- *Marine Radar Jamming*: Radar jamming refers that attackers deliberately sending radio frequency signals that can make receivers full of noise to interfere with the operation of radars. When radio frequency signals cover radar signals, radar operators are not able to read the radar data. In general, radar jamming is categorized into two types: mechanical and electronic as described below [8].

  - *Mechanical Jamming*: Through using devices such as chaffs, decoys, or reflectors, attackers can reflect or re-reflect radar signals back to transmitting sources. In this case, radar operators will receive false target signals. For example, a corner reflector can be usually installed on a decoy to confuse radar operators to make them look like a real plane.
  - *Electronic Jamming*: Electronic jamming is a form of electronic warfare. Attackers send jamming signals to enemy radars by utilizing jammers. In this case, receivers are blocked by highly concentrated jamming signals. Repeater and noise techniques are currently two main techniques. There are three categories of noise jamming: spot, sweep, and barrage described as follows [9]:

    * *Spot Jamming*: Spot jamming happens when a jammer concentrates all its power on a single particular frequency. Since a single particular frequency is affected by spot jamming, a radar display cannot show a target. This threat is only for a single frequency radar.
    * *Sweep Jamming*: Sweep jamming, which is an upgrade version of spot jamming, improves spot jamming because sweep jamming can attack multiple frequencies. However, sweep jamming cannot attack all frequencies at the same time.
    * *Barrage Jamming*: Barrage jamming, which is a perfected version of spot jamming, can attack multiple frequencies simultaneously. However, barrage jamming consumes too much power to perform this operation, so the effect of jamming may be limited.

- *Marine Radar Spoofing*: A radar determines the distance of an object by measuring the time it takes for a signal to return to the radar. If an attacker changes the return time of a radar signal, a radar display will show an object in the wrong position. In general, radar operators cannot identify spoofing at radar displays. The consequences of spoofing may lead to ship' collisions, or ships entering an unknown sea area and cause dangers.

7.4. Security Issues of Localization in Smart Ships

A localization system is a mechanism that can track, identify, and monitor the status and positions of ships in the ocean. Localization systems are essential systems that provide the safety of ships. For example, localization systems can monitor the positions of ships to prevent collisions and traffic jams when ships enter ports. Meanwhile, localization systems can also help ships fix in a particular position in the ocean without laying anchors. In this section, we will introduce various security issues of localization systems in detail.

7.4.1. *Automatic Identification System*

Automatic Identification System (AIS), one kind of navigation aid systems, helps boat captains to be aware of locations of other ships and to receive AIS signals from nearby ships. AIS provides many sailing information, such as course, ground speed, unique identification, and GPS coordinates, which can appear on an Electronic Chart Display and Information System (ECDIS). Meanwhile, compared with low accurate radar technology, AIS is more valuable for search-and-rescue [10]. As with other navigational equipment, AIS also has issues with security and privacy. Figure 7.1 shows that AIS-related security issues which are categorized into four types: spoofing, hijacking, failure, and impersonating.

- *Ship Spoofing*: Attackers counterfeit the AIS signal of someone else's vessel or craft the AIS signal of virtual non existent vessels to commit vessel spoofing on a broad sea. On the ocean, identifying tracks and sailings of ships is very challenging because attackers can assign virtual vessels fraudulent dynamic information and fake static information, such as vessel' identities, types, positions, courses, destinations, etc. Ship spoofing can be applied to a broad range of malicious scenarios, such as spoofing a rescue aircraft to search outside of the rescue range. Sometimes, spoofing ships into the waters of other countries may cause conflicts or wars.
- *Aids to Navigation Systems*: Aid to Navigation (ATON), which is also called navigational aid, is usually utilized to assist vessel transportations or to warn operators about potential hazards, such as low

Macro Category	Threat	Software Based	Radio Frequency Based
Spoofing	Ship Spoofing	√	√
	Aids to Navigation Systems	√	√
	Collision Spoofing	×	√
	AIS-Search and Rescue Transmitter	√	√
	False Weather Forecasting	×	√
Hijacking	AIS Hijacking	√	√
Failure	Human Errors	×	×
Impersonating	Impersonating Marine Authorities	×	√

Fig. 7.1.   AIS-related threats

tides [11]. ATON commonly includes beacons, lighthouses, lights, or buoys. Since AIS transmitters can be appended to ATON, ATON is usually classified into three types: real AIS ATON, synthetic AIS ATON, and virtual AIS ATON [12]. Although ATON effectively improves the safety of vessels, ATON spoofing is a potential threat to the security of vessels. For example, attackers can install fake buoys to maliciously lead vessels to low water. Another example is that attackers can broadcast fraudulent information, such as one or more buoys existing at the port entrance to lure operators to make wrong maneuvers.

- *Collision Spoofing*: Collision avoidance, which is one of the significant applications in AIS, enhances navigational efficiency, and improves operations of vessels [10]. As an assistant function to Officer on Watch (OOW), the AIS system can automatically detect expected collisions through a function of Closest Point of Approach (CPA) [13]. When the CPA system triggers a collision alert, vessels will automatically change course to avoid an accident [10]. A potential security issue is that when tricking a vessel collide with a targeted vessel, a collision alert on the victim ship may cause the vessel to veer off course into waters of other countries or strands vessels in the shallow water.

- *AIS-Search And Rescue Transmitter (AIS-SART) Spoofing*: Search And Rescue Transponders (SARTs), which are widely used in search and rescue operations, are self-contained radio devices that transmit AIS messages, such as vessel' positions and static information in distress

[14]. AIS-SART will be activated automatically when it is in contact with water and will send a distress message automatically with the location to AIS stations via radio beacon. A potential security issue is that attackers send false distress messages containing incorrect coordinates to AIS stations and trigger SART alerts [10]. In this case, attackers lure rescue vessels into unsafe sea space or the waters of other countries.

- *False Weather Forecasting*: To reflect the changing situations at sea, dynamic information of AIS updates in intervals between two seconds and three minutes [15]. When a fake weather forecast is announced to a targeted vessel by attackers, victims may be navigated into typhoons and storms.

- *AIS Hijacking*: AIS hijacking involves modifying any existing information from AIS stations. For example, these modifications contain vessel' locations, courses, speeds, types of cargo, etc. An example of AIS hijacking is that attackers override original signals of AIS through broadcasting higher power signals [10]. In this scenario, attackers can arbitrarily alter original AIS messages. Another example is that attackers can modify AIS messages through illegal eavesdropping attacks [10].

- *Human Errors*: The coded messages of AIS contain not only the information of vessels but also Maritime Mobile Service Identity (MMSI), which similar to a telephone number, is a series of nine digital numbers to identify each vessel [16]. Each operator must manually type the AIS message into the transponder. Manual input may cause a lot of room for errors, intentionally or unintentionally. To save time, irresponsible operators usually use a simple, meaningless MMSI, such as MMSI 531000000 [16]. Wrong MMSI codes may cause multiple vessels to simultaneously broadcast the identical MMSI number, which cannot recognize disparate vessels. When vessels send distress radio messages, identifying tracks is very challenging. Therefore, an incorrect MMSI will lead to delayed rescue time.

- *Impersonating Marine Authorities*: Attackers impersonate marine authorities to control and keep the entire AIS transmission "address space" for preventing all stations with coverage from communicating with each other [10]. Stations include ships, ATON, and AIS gateways that are utilized in traffic monitoring. Therefore, attackers can ban AISs on a large scale.

7.4.2. *Dynamic Positioning System*

For many ships on long-distance voyages or offshore work, due to uncertain marine environment factors, such as winds, waves, or currents, ships are difficult to maintain fixed positions or planned courses. Dynamic Positioning (DP) is an automatic control system of positions and courses of ships. When ships deviate from their fixed positions or planned courses, DP systems can automatically maintain positions and courses of ships by using ship' propellers and thrusters. Although DP systems can track, identify, and monitor the status of ships, DP systems still have the following security issues:

- *Human Errors*: Human errors in DP systems can lead to severe accidents, such as collision or property damage. There is a real case that happened in the United States where an operator of the DP system in a drillship inadvertently double-clicked a manual button to stop the automatic position mode while passing through the console [17]. After the mistake occured, an operator immediately repositioned this ship to restore its original position and avoid a collision. Fortunately, crews and the ship were not harmed. However, if this ship's drills and other equipment had collided with the undersea reef, the ship or equipment would have been damaged.
- *Hardware and Software Failure*: Due to hardware and software failure of PD systems, ships may lose positions or courses. Potential failures of PD systems may include thruster, generator, power bus, control computer, position reference system, and reference system [18]. For example, for offshore exploration and gas drilling, hardware and software failure of DP systems may cause ships to deviate from the original position, resulting in equipment damage.

7.4.3. *Long Range Identification and Tracking*

The Long Range Identification and Tracking (LRIT) is a tracking and identification system for ships around the world. LRIT can track the positions, monitor the information, and identify the status of ships. Meanwhile, LRIT utilizes radio links to establish communication between ship' transponders and stations on land. However, there are several severe deficiencies in the security of the LRIT system. We will analyze these security issues in detail next.

- *Eavesdropping*: LRIT systems use radio links to transmit data between ship' transponders and stations on land. When channels are opened between ships and shore, messages and data of transmission are very susceptible to eavesdropping [19]. If data of transmission is plain text that is not encrypted, unauthorized stations easily receive the plain text.

7.5. Security Issues of Path Planning in Smart Ships

A path planning system is a major component of a ships' autonomy systems, because it can help ships avoid navigation dangers. However, path planning system vulnerabilities may cause potential security risks as described next.

7.5.1. *Electronic Chart Display and Information System*

An Electronic Chart Display and Information System (ECDIS) is another great technological revolution in vessel navigation after radars have been widely implemented. It not only provides all kinds of information about navigation, but also it effectively prevents varieties of dangers. Critical functionalities of ECDIS include automatic path planning, path monitoring, path update management, etc., [20]. Although ECDIS has powerful functionalities, it still has many issues of security, such as backup configuration issues, third party services, and operating system vulnerabilities [21]. These security vulnerabilities are as follows:

- *ECDIS Backup Configuration Issues*: Since each ECDIS station is installed with identical underlying software and hardware, ECDIS stations may suffer the same level of vulnerability to a single threat [21]. The same vulnerability level of ECDIS stations provides an ideal environment to assign malicious executive codes through the integration network or a portable storage device.
- *Operating System Vulnerabilities*: ECDIS's operating system has critical vulnerabilities, which are related to Server Message Block (SMB) version 1 [21]. In this case, ECDIS's operating system needs to install the latest security patches immediately. The SMB service vulnerability also has a dangerous impact on the maritime industry. For example, NotPetya, which is a malicious blackmail software program, spreads rapidly around the world through utilizing vulnerabilities of the SMB version 1 [21]. Meanwhile, in the case of ECDIS backup arrangements,

infection of one single ECDIS station very likely leads to infection or fault of other ECDIS stations in the network [21].

- *Third-Party Services Vulnerabilities*: Third-party services such as web server applications and remote desktop control applications have many risky vulnerabilities that cause ECDIS crashing [21]. For example, outdated Apache web servers on the ECDIS might lead to unauthenticated and remote attackers to gain sensitive information about vessels, launch a Denial of Service (DoS) attack, and execute malicious code.

7.6. Privacy Issues of Navigation, Localization, Path Planning

No matter what systems of navigation, localization, or path planning are used by ship operators, the privacy of ship information may disclose ship' size, position, course, etc. Information on ships may reveal directions or the current status of ships. If ship' information is maliciously collected or monitored without permission, the safety of crew and the ship may be potentially in danger. These privacy issues of ships include:

- *Tracking Locations of Navigation*: Navigation systems of ships provide accurate geographical locations and real-time updated navigation systems for ensuring the safety and stability of vessels. However, GNSS operating systems could record vessel' positions or other locations that vessels visit often. Organizations that collect and store data may spy on locations of vessels or sell a specific location or course of a ship to a third party for profit.
- *Identifying Information of Ships*: Systems of navigation, localization, and path planning include various sensitive information of ships, such as position, status, course, size, etc. If information about ships is not encrypted, unauthorized users or stations may access information during storage and exchange.

7.7. Security and Privacy Countermeasures

Although systems of navigation, localization, and path planning have many security and privacy issues, we found some solutions that can prevent or mitigate these issues. Some countermeasures are as follows:

- *Mitigating Jamming Attacks of Marine Radar*: Radar operators can constantly alter radar frequencies to limit the effectiveness of most types of jamming. However, modern jammers can predict and track frequency changes. Therefore, the possibility of resisting jamming is decided by the randomness of frequency changes. For example, if radar frequency changes are more random, the possibility of resisting jamming is more likely.
- *Time Difference of Arrival of AIS*: Most AIS spoofings, such as ships, ATON, and collision, rely on broadcasting false locations. To distinguish whether signals are real or a spoof, the best solution is to find real locations of transmission AIS signals. CRFS RFeye Node, which is a spectrum monitoring node, can geolocate real locations of transmission AIS signals by using Time Difference Of Arrival (TDOA) [22]. When actual locations of transmission AIS signals are determined, operators of AIS can compare with locations given by AIS messages. If two locations have a significant discrepancy, attackers are carrying out spoofing.
- *Key Performance Indicators (KPIs) of GNSS Receivers*: Monitoring KPIs of GNSS receivers prevent simple spoofing attacks. For example, operators can monitor clock jumps, compare the difference between code and carrier measurements, or check unusual signal-to-noise density ratios.
- *Mitigating Spoofing and Jamming Attacks of GNAA*: Array antennas, such as Controlled Reception Pattern Antennas (CRPA), can mitigate jamming and spoofing of GNSS. Array antennas, which are a group of connected antennas, transmit or receive radio waves as a single antenna. Since the radio wave of each antenna is superimposed, array antennas enhance the power of radiation and mitigate jamming in desired directions. Therefore, array antennas effectively increase the reliability of the communication of GNAA.

7.8. Conclusion

In this chapter, we gave a comprehensive analysis of security and privacy issues of navigation, localization, as well as path planning. Firstly, we introduced concepts of navigation, localization, and path planning. Secondly, we analyzed their security and privacy issues. Finally, we listed several solutions to solve their issues. These countermeasures can prevent or mitigate security and privacy issues. As more and more technological components are used on smart ships, it is worthwhile to conduct more research on the security and privacy aspects of these new components.

References

1. J. Wang, Y. Xiao, T. Li, and C. L. P. Chen, "A survey of technologies for unmanned merchant ships," *IEEE Access*, vol. 8, pp. 224 461–224 486, 2020.
2. K. Chopra, "The history of ships: Ancient maritime world," https://www. marineinsight.com/maritime-history/the-history-of-ships-ancient-maritim e-world/, Accessed Oct 10, 2019.
3. R. Pros-Wellenhof, K. Legat, and M. Wieser, *Navigation: Principles of Positioning and Guidances.* Springer-Verlag, New York, 2003.
4. M. Kaushik, "What is a dynamic positioning ship?" https://www. marineinsight.com/types-of-ships/what-is-a-dynamic-positioning-ship/, Accessed May 20, 2020.
5. Intertanko, "Jamming and spoofing of global navigation satellite systems (gnss)," https://www.maritimeglobalsecurity.org/media/1043/2019-jammi ng-spoofing-of-gnss.pdf/, Accessed May 20, 2020.
6. R. Morales-Ferre, P. Richter, E. Falletti, A. de la Fuente, and E. S. Lohan, "A survey on coping with intentional interference in satellite navigation for manned and unmanned aircraft," *IEEE Communications Surveys & Tutorials*, vol. 22, pp. 249–291, 2020.
7. M.-H. Golbon-Haghighi and G. Zhang, "Detection of ground clutter for dual-polarization weather radar using a novel 3d discriminant function," *Journal of Atmospheric and Oceanic Technology*, vol. 36, pp. 1285–1296, 2019.
8. Wikipedia, "Radar jamming and deception," https://en.wikipedia.org/wik i/Radar_jamming_and_deception/, Accessed May 20, 2020.
9. F. A. Butt and M. Jalil, "An overview of electronic warfare in radar systems," in *Proc. Int. Conf. Technol. Adv. Elect. Electron. Comput. Eng. (TAEECE)*, 2013, pp. 213–217.
10. M. Balduzzi, A. Pasta, and K. Wilhoit, "A security evaluation of ais automated identification system," in *Proceedings of the 30th Annual Computer Security Applications Conference, (ACSAC 2014), ACM, New York, NY, USA*, 2014, pp. 436–445.
11. Wikipedia, "Navigational aid," https://en.wikipedia.org/wiki/Navigational _aid/, Accessed May 20, 2020.
12. Office of Coast Survey, "Portrayal of AIS aids to navigation," https: //nauticalcharts.noaa.gov/publications/portrayal-of-ais-aids-to- navigation.html/, Accessed May 20, 2020.
13. S. Bhattacharjee, "Automatic identification system (ais): Integrating and identifying marine communication channels," https://www.marineinsight. com/marine-navigation/automatic-identification-system-ais-integrating-a nd-identifying-\\marine-communication-channels/, Accessed May 20, 2020.

14. M. Balduzzi, K. Wilhoit, and A. Pasta, "A security evaluation of ais," http://citeseerx.ist.psu.edu/viewdoc/download?doi=10.1.1.685.2706&rep= rep1&type=pdf, pp. 1–33, 2014, accessed Oct 10, 2019.

15. I. Harre, "AIS adding new quality to vts systems," *The Journal of Navigation*, vol. 53, pp. 527–539, 2000.

16. K. Cutlip, "Spoofing: One identity shared by multiple vessels," https:// globalfishingwatch.org/data/spoofing-one-identity-shared-by-multiple-vessels/, Accessed May 20, 2020.

17. Nopsema, "Dynamic position control systems must be tolerant to human error," https://www.nopsema.gov.au/safety/safety-resources/dynamic-position-control-systems-must-be-tolerant-to-human-error/, Accessed May 20, 2020.

18. Wikipedia, "Dynamic positioning," https://en.wikipedia.org/wiki/Dynamic_ positioning#Dynamic_positioning_alarm_and_runout_response_for_bell_diver s/, Accessed May 20, 2020.

19. R. Niski, M. Waraksa, and J. Zurek, "Information security in the LRIT system," in *2008 1st International Conference on Information Technology, Gdansk, Poland*, 2008, pp. 1–4.

20. M. Caprolu, R. D. Pietro, S. Raponi, S. Sciancalepore, and P. Tedeschi, "Vessels cybersecurity: Issues challenges and the road ahead," *IEEE Communications Magazine*, vol. 58, pp. 90–96, 2020.

21. B. Svilicic, D. Brcic, S. Zuskin, and D. Kalebic., "Raising awareness on cyber security of ECDIS," *The International Journal on Marine Navigation and Safety of Sea Transportation*, vol. 13, pp. 231–236, 2019.

22. CRFS, "AIS spoofing detection with TDOA," https://www.crfs.com/blog/ ais-spoofing-detection-with-tdoa/, Accessed May 20, 2020.

PART 4

AUV and Communication Technologies in Smart Ships

Chapter 8

Smart AUV-Assisted Localization and Time Synchronization of Underwater Acoustic Devices

Zijun Gong[1], Fan Jiang[2], Ruoyu Su[3] and Cheng Li[4,*]

[1] *Department of Electrical and Computer Engineering, University of Waterloo, 200 University Avenue W, Waterloo, ON, N2L 3G1, Canada (E-mail: zijun.gong@uwaterloo.ca).*

[2] *Department of Electrical Engineering, Chalmers University of Technology, SE-412 96, Göteborg, Sweden (e-mail: fan.jiang@chalmers.se).*

[3] *School of Internet of Things, Nanjing University of Posts & Telecommunications, 210023, Nanjing, China (e-mail: suruoyu@njupt.edu.cn).*

[4] *Faculty of Engineering and Applied Science, Memorial University of Newfoundland, St. John's, NL, A1B 3X5, Canada (E-mail: licheng@mun.ca).*
* *Cheng Li is the corresponding author.*

8.1. Introduction

With the explosive growth of human appetite for ocean exploration, the underwater acoustic sensor networks are playing more and more important roles in resource exploration, natural disaster forewarning, and surveillance of target water areas. For coastal guards, the acoustic sensor networks also provide solutions for the detection and tracking of illegal fishing and smuggling boats. For these purposes, an efficient localization and synchronization architecture for these sensor networks is very important. The conventional architecture based on fixed anchors has poor scalability, and it is highly dependent on anchor density. For very large areas, it is almost impossible to provide good coverage at an acceptable cost. Besides, the large propagation delay and variant velocity of acoustic signals add more challenges.

Considering the limits of the conventional architecture, many researchers propose to employ smart AUVs as mobile anchors to replace the fixed ones [1]. By moving close to the sensors, the AUVs can also collect the data through optical communications, which supports very high data rates. However, the AUVs must localize the sensors before that. To achieve this goal, the AUVs travel on the predefined trajectories, and broadcast beacon signals periodically. Any acoustic device in the communication range of the AUVs can receive the beacon signals and localize themselves through trilateration. This system can also work in silent mode. To be specific, the target devices generate acoustic signals, while the AUVs stay silent. From the received signals, the AUVs can localize and track the target devices. This new architecture has many potential applications, such as wreck salvage. The safe box of the sunk ship or plane will keep broadcasting beacon signals, and we can deploy multiple AUVs in the search area to facilitate the rescue process. Nowadays, some AUVs are solar-powered and suitable for long-endurance missions [2]. By employing AUVs, the number (or density, equivalently) of anchors can be boosted at negligible cost, leading to much-improved accuracy. Also, the AUVs can provide very good coverage by moving around in the area of interest.

In this chapter, we will talk about the application of smart AUVs in the underwater localization and synchronization of acoustic devices. The following topics will be covered:

1. ToA-based joint localization and synchronization of underwater acoustic sensors with the assistance of smart AUVs;
2. Doppler-based localization of UADs with the assistance of smart AUVs;
3. Application of the linear-frequency-modulated signals for joint range-velocity estimation.

These three topics will be discussed in the following three sections, and the last section is a brief summary.

8.2. ToA-Based Joint Localization and Time Synchronization

For underwater acoustic sensor networks, location information and time synchronization of all nodes are critical or at least of great value in many applications. However, the localization and time synchronization of underwater sensors can be quite challenging for the following reasons. First, acoustic communication systems are generally employed over the electromagnetic ones for underwater applications, which introduces long propagation delay. Besides, underwater sound speed is not constant due to the heterogeneous salinity, density, and temperature of seawater, leading to the well-known stratification effect [3]. Although an algorithm has been proposed to compensate for the stratification effect in [4], it is computationally intense. In [5], the authors showed that the stratification effect can be ignored by viewing the average acoustic speed as an extra unknown in the localization and synchronization algorithm.

For both terrestrial and underwater wireless sensor networks, time synchronization is generally achieved through message exchange between location-aware fixed anchors (buoys) and target sensors, while localization can be based on ToA (Time of Arrival), TDoA (Time Difference of Arrival) [6], RSS (Received Signal Strength) [7, 8], and DoA (Direction of Arrival) [9]. To the best of the authors' knowledge, TDoA- and ToA-based methods are still the most common choice, because they can serve both time synchronization and localization purposes. Besides, the estimation error of TDoA/ToA is reported to be at the level of milliseconds (ms) [10], leading to high-precision ranging results. Based on the TDoA/ToA measurements, the ML (Maximum Likelihood) estimator is reported to achieve the Cramér-Rao Lower Bound (CRLB) at the cost of expensive computation; therefore, a more efficient sub-optimal least square (LS) estimator is presented in [11]. Besides, the uncertainty in the anchor's location information is also considered.

For the localization and time synchronization system based on fixed anchors, a huge number of anchors are required. Besides, because the anchors are fixed, their coverage will be quite limited. Therefore, the authors of [1] proposed to use an AUV to serve as a mobile anchor. The AUV travels on predefined trajectories and broadcasts beacon signals periodically, and by receiving them, all the nodes can localize themselves. In [5], the possibility of AUV-assisted time synchronization is also discussed and the Sequential Time-Synchronization and Localization (STSL) algorithm is proposed, in which time synchronization and localization are conducted iteratively. This method has two problems: first, the extra estimation error is introduced by assuming the AUV is static in a short period of time; second, according to the simulation results, the iteration process converges slowly.

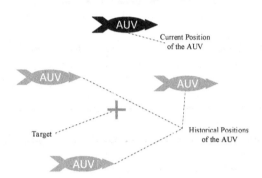

Fig. 8.1.   Illustration of the AUV-assisted localization and time synchronization system

In this section, the system model of AUV-assisted localization and time synchronization will be presented, based on TDoA measurements. A two-phase linear algorithm for joint time synchronization and localization will be introduced [12]. The CRLB of the whole system will be investigated, and some numerical evaluations will be provided.

8.2.1. *System Model*

The AUVs generally have on-board navigation systems, such as GPS chips and inertial navigation sensors. An AUV can initialize its position information with the on-board GPS chip before diving. Then, it will dive and navigate itself through predefined trajectories with an inertial sensor. To avoid error accumulation in inertial sensors, the AUVs can surface periodically to update their position information through GPS.

With an AUV, Fig. 8.1 shows the basic idea of how the localization and time synchronization can be achieved for a target node. The AUV travels on the predefined trajectory and moves from one position to another. At each position, the AUV broadcasts its 3-dimensional (3D) coordinates, serving as a virtual anchor. The target node will receive the beacon messages from the AUV periodically.

To achieve 3D localization, the trajectory of the AUV cannot be on a plane. Therefore, we assume that the AUV moves with fixed direction and velocity for K time slots, and then it turns to a randomly selected direction and moves forward at the new direction with fixed velocity for another K time slot. During this process, the AUV periodically broadcasts beacon signals, including its real-time location information and transmitting time of packets. This process will continue until we have N straight lines on AUV's trajectory.

On a specific line, assume that the location information and the transmitting time of the k-th period is $\mathbf{x}_k$ and t_k, respectively. When this packet is received by a target sensor at time r_k, the measured propagation delay will be

$$T_k = s \cdot r_k + o - t_k = \overline{T}_k + n_k, \tag{8.1}$$

where s and o denote the clock skew and offset of the clock on target sensor, with respect to the clock on AUV. $\overline{T}_k$ is the accurate propagation time and n_k is the overall timing error. Both r_k and t_k are contaminated by zero-mean Gaussian noise with an identical variance of σ_t^2. Therefore, the total timing error will be a zero-mean Gaussian variable with a variance of $(s^2 + 1)\sigma_t^2$. On the other hand, based on the ToA measurement, we have

$$\|\mathbf{x} - \mathbf{x}_k\| = \overline{T}_k \cdot c, \tag{8.2}$$

where $\mathbf{x}$ is the 3D coordinate of the target, c is the average underwater sound speed, and $\| \cdot \|$ indicates the Euclidean norm of an arbitrary vector.

Suppose the trajectory contains N straight lines. After taking K measurements on each of them, we have NK nonlinear equations. The maximum likelihood estimate of the unknowns is thus given as

$$\hat{\boldsymbol{\theta}}_{ml} = \arg \min_{\boldsymbol{\theta}} \sum_{n=1}^{N} \sum_{k=1}^{K} |f_{n,k}(\boldsymbol{\theta})|^2, \tag{8.3}$$

where $\boldsymbol{\theta} \triangleq [\mathbf{x}^T, o, c, s]^T$ contain the six unknowns, and $f_{n,k}(\boldsymbol{\theta})$ is the nonlinear equation obtained from the measurement in the k-th period on the n-th line.

Generally, these nonlinear equations can be iteratively solved by many algorithms (e.g., Newton's method). However, this is a nonconvex problem, and the iteration algorithm may be stuck at local optimums if we employ random initializations of the unknowns [5]. To solve this problem, a two-phase algorithm is presented in [12], which proves to achieve the CRLB.

8.2.2. *Two-Phase Low-Complexity Algorithm*

In the first phase, the relative clock skew is ignored because it is generally very small. Then, the nonlinear equations are transformed into linear ones, and the LS algorithm is employed to obtain coarse time synchronization and localization results. In the second phase, the coarse estimation is refined by another LS estimator.

8.2.2.1. *Phase I: initial synchronization and localization results*

Generally, clock skew is smaller than 200 ppm [13–15], and the corresponding s lies in $[1 - 2\text{E-4}, 1 + 2\text{E-4}]$, which is very close to 1. Therefore, we can replace it with $\hat{s} = 1$ to roughly estimate the other unknowns. For brevity, define $q_k \triangleq \hat{s} \cdot r_k - t_k$, square both sides of Eq. (8.2), and we will have K equations of the following form:

$$\|\mathbf{x} - \mathbf{x}_k\|^2 - q_k^2 c^2 - o^2 c^2 - 2q_k o c^2 = e_k, \tag{8.4}$$

where e_k is the overall error caused by timing error and the uncertainty in $\hat{s}$. Then, after subtracting the first equation from the k-th one, we can obtain

$$
\begin{aligned}
(\mathbf{x}_1 - \mathbf{x}_k)^T \mathbf{x} + (q_1^2 - q_k^2)/2 \cdot c^2 + (q_1 - q_k) \cdot c^2 o \\
= (\|\mathbf{x}_1\|^2 - \|\mathbf{x}_k\|^2)/2 + (e_k - e_1)/2 \text{ (for } k \neq 1).
\end{aligned}
\tag{8.5}
$$

If we define the unknown vector $\mathbf{p}$ as

$$\mathbf{p} = \left[\mathbf{x}^T, c^2, c^2 o\right]^T, \tag{8.6}$$

we can clearly see that Eq. (8.5) is a linear equation of $\mathbf{p}$. In total, we have $K - 1$ equations in this form. Then, by stacking the equations, we can employ the LS algorithm to estimate $\mathbf{p}$, and we can get the positioning result as $\hat{\mathbf{x}}$. The clock offset and average underwater sound speed are estimated as $\hat{o}$ and $\hat{c}$, respectively. Therefore, a coarse estimation of $\boldsymbol{\theta}$ can be obtained as $\hat{\boldsymbol{\theta}} = [\hat{\mathbf{x}}^T, \hat{o}, \hat{c}, \hat{s}]^T$, where $\hat{s}$ is equal to 1.

To guarantee the positioning and timing accuracy, the trajectory of the AUV should be carefully designed to make sure that Eq. (8.4) is not ill-conditioned. Because of the space limit, we will not discuss this issue here, and the interested readers are referred to [16] and [17].

8.2.2.2. *Phase II: refined synchronization and localization results*

After the coarse estimation in Phase I, $\hat{\boldsymbol{\theta}}$ should be reasonably close to $\boldsymbol{\theta}$, which leads to the following approximation

$$\mathbf{f}(\boldsymbol{\theta}) - \mathbf{f}(\hat{\boldsymbol{\theta}}) \approx \frac{\partial \mathbf{f}}{\partial \boldsymbol{\theta}} \cdot (\boldsymbol{\theta} - \hat{\boldsymbol{\theta}}), \tag{8.7}$$

where $\mathbf{f}(\boldsymbol{\theta})$ is a vector containing all the observed functions. Based on this approximation, $\boldsymbol{\delta}_\theta = \boldsymbol{\theta} - \hat{\boldsymbol{\theta}}$ can be estimated to refine the estimated results. As we known, ToA measurement errors are generally very small (at the level of ms [10]), which means we have $\mathbf{f}(\boldsymbol{\theta}) \approx \mathbf{0}_{KN \times 1}$. Moreover, the Jacobian matrix can be replaced by

$$\mathbf{R} = \left. \frac{\partial \mathbf{f}}{\partial \boldsymbol{\theta}} \right|_{\boldsymbol{\theta}=\hat{\boldsymbol{\theta}}}. \tag{8.8}$$

After the replacement, the approximation in Eq. (8.7) can be revised as

$$-\mathbf{f}(\hat{\boldsymbol{\theta}}) \approx \mathbf{R}\boldsymbol{\delta_\theta}. \tag{8.9}$$

Then, we can estimate the bias of coarse estimation in Phase I as

$$\hat{\boldsymbol{\delta_\theta}} = -(\mathbf{R}^T\mathbf{R})^{-1}\mathbf{R}^T\mathbf{f}(\hat{\boldsymbol{\theta}}), \tag{8.10}$$

and the time synchronization and localization results can be refined as

$$\hat{\boldsymbol{\theta}}_r = \hat{\boldsymbol{\theta}} + \hat{\boldsymbol{\delta_\theta}}. \tag{8.11}$$

It should be noted that this process can be repeated to get more accurate results. Besides, we only deal with localization and time synchronization in this paper, and when it comes to tracking, the particle filter is a popular tool to further improve accuracy [18–20].

8.2.3. *CRLB Analyses*

In Eq. (8.10), the noise in ToA measurements will cause estimation error, and similar to the works in [21] and [22], the covariance matrix of estimation error can be approximated by

$$\text{cov}\{\hat{\boldsymbol{\theta}}\} = (s^2 + 1)\sigma_t^2(\mathbf{R}^T\mathbf{R})^{-1}. \tag{8.12}$$

Now an immediate question is whether can we further improve the system performance by developing more sophisticated algorithms. To answer this question, we need to derive the CRLB of the system.

In Eq. (8.5), we can see that overall timing error follows zero-mean Gaussian distribution, with a variance of $(s^2 + 1)\sigma_t^2$. We assume all the timing errors are independent and identically distributed, and the joint probability density function of the observations is

$$f_{\mathbf{r},\mathbf{t}}(\mathbf{r}, \mathbf{t}|\boldsymbol{\theta}) = (|2\pi\boldsymbol{\Sigma}|)^{-1/2} \cdot \exp\left\{-\frac{1}{2}\mathbf{f}(\boldsymbol{\theta})^T\boldsymbol{\Sigma}^{-1}\mathbf{f}(\boldsymbol{\theta})\right\}, \tag{8.13}$$

where $\mathbf{r}$ is the time of receiving vector, and $\mathbf{t}$ is the time of transmission vector. The covariance matrix will be $\boldsymbol{\Sigma} = (s^2+1)\cdot\sigma_t^2\mathbf{I}_{NK}$, and $|\cdot|$ represents the determinant of an arbitrary square matrix. The natural logarithm of $f_{\mathbf{r},\mathbf{t}}(\mathbf{r},\mathbf{t}|\boldsymbol{\theta})$ is

$$l(\boldsymbol{\theta}) = -\frac{1}{2}\left(\ln|2\pi\boldsymbol{\Sigma}| + \mathbf{f}(\boldsymbol{\theta})^T\boldsymbol{\Sigma}^{-1}\mathbf{f}(\boldsymbol{\theta})\right). \qquad (8.14)$$

Define $\mathbf{F}$ as the Fisher Information Matrix (FIM), and the (m,n)-th element of $\mathbf{F}$ will be

$$[\mathbf{F}]_{m,n} = E\left\{\frac{\partial l}{\partial \theta_m}\cdot\frac{\partial l}{\partial \theta_n}\right\}, \qquad (8.15)$$

where θ_m and θ_n are the m-th and n-th elements of $\boldsymbol{\theta}$, respectively. The FIM can be derived in [12] as

$$\mathbf{F} = \frac{1}{(s^2+1)\sigma_t^2}\left(\mathbf{R}_o^T\mathbf{R}_o + \begin{bmatrix}\mathbf{0}_{5\times 5} & \mathbf{0}_{5\times 1} \\ \mathbf{0}_{1\times 5} & \frac{N^2K^2s^2\sigma_t^2}{s^2+1}\end{bmatrix}\right), \qquad (8.16)$$

where $\mathbf{R}_o$ is the Jacobian matrix evaluated at $\boldsymbol{\theta}$.

Given that the estimation results are very close to the actual values, we have $\mathbf{R}\approx\mathbf{R}_o$. Besides, compared with the elements in $\mathbf{R}_o^T\mathbf{R}_o$, the nonzero element at the right bottom of the additional matrix is negligible because σ_t^2 is very small. Thus, we have $\mathrm{cov}\{\hat{\boldsymbol{\theta}}\}\approx\mathbf{F}^{-1}$, and the estimation error of the proposed algorithm is very close to CRLB. Now we can answer our previous question, and we know that this low-complexity algorithm can already achieve the performance bound.

We can easily see that the overall computational complexity increases linearly with the number of observations. The algorithm in [11] has a computational complexity of the same order but several times larger, because it contains nine unknowns.

8.2.4. *Numerical Evaluation*

In this section, we will conduct Monte Carlo simulation for one hundred thousand times to evaluate system performance and verify our analysis in previous sections. The average underwater sound speed is uniformly distributed between 1420 and 1560 m/s as in [23]. The broadcast interval is set as 5 seconds, and AUV velocity is chosen between 1.5 to 2.5 m/s. N and K are equal to 4 and 50, respectively. The standard deviation of timing error varies from 0.1 to 1.6 ms. For visual convenience, σ_t^2 is transformed to decibel according to $\sigma_t^2/\mathrm{dB} = 10\lg(1000\sigma_t)^2 = 10\lg\sigma_t^2 + 60$, where $\lg(\cdot)$ represents the base-10 logarithm. Clock skew is set as $1 + 1\mathrm{E} - 4$ (or 100 ppm).

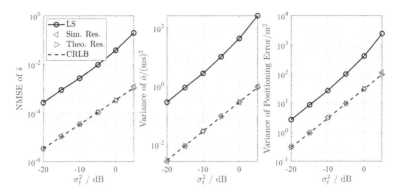

Fig. 8.2. Localization and synchronization errors of the proposed algorithm and the LS algorithm in [11]. "Sim. Res." and "Theo. Res." are short for "Simulation Results" and "Theoretical Results" (in Eq. (8.12)).

The Normalized Mean Squared Error (NMSE) of $\hat{s}$ is defined as

$$\text{NMSE of } \hat{s} = \left(\frac{\hat{s}-s}{s-1}\right)^2, \tag{8.17}$$

which indicates the relative estimation error with respect to the fractional part of s.

The simulation results are shown in Fig. 8.2, including the estimation error of the target sensor's position, clock skew, clock offset, and the average underwater sound speed. In comparison, the LS algorithm proposed in [11] is also simulated. As we can see, the presented method outperforms the LS algorithm in [11]. Besides, the results show that the performance analysis in Eq. (8.12) is very accurate in the simulation range of σ_t^2. Moreover, it is clear that this algorithm has a very good approach to CRLB. As we can see, the discrepancy between CRLB and simulation results increases with the growth of σ_t^2, as is suggested by Eq. (8.16).

8.3. Doppler-Based Localization

From the previous section, we can see that the ToA-based localization technique for UADs has great performance. However, the ToA measurements generally demand two-way message exchange and are not readily available. As an alternative, we will investigate the possibility of employing Doppler shift measurements for underwater localization of acoustic devices in this section. For underwater applications, the Doppler shift is generally employed to estimate the target's velocity. However, we will show that the Doppler shift measurements also contain the target's location information.

Again, an AUV moves around in the area of interest and serves as a mobile anchor. This system can work in both proactive and passive modes. In the proactive mode, the AUV broadcasts its location information and a sinusoidal wave periodically. The target devices can localize themselves by receiving the signals from the AUV. In the passive mode, the AUV stays silent and receives sinusoid signals from the target devices. Based on the received signal, the AUV can estimate the location of the targets. Compared with the ToA- or TDoA-based systems, the Doppler-based system has many advantages. First, localization accuracy can be boosted at a very low cost. For example, suppose we need to increase the localization accuracy by one order of magnitude. For the ToA- or TDoA-based methods, the AUV needs to broadcast one hundred times faster (or longer). For the Doppler-based method, we just need to sample the sinusoidal wave 4.5 ($\sqrt[3]{100} \approx 4.5$) times longer, as will be shown in later discussions. Second, the ToA and TDoA measurements are not always available for various underwater localization applications, because our target does not necessarily have on-board communication modules. For example, assume that we want to track some moving objects, such as sharks and whales. We only need to attach a very simple tag that can generate sinusoidal acoustic waves for the Doppler-based localization system. However, for the ToA- or TDoA-based methods, we will need to install a much more complicated device for bidirectional communications.

8.3.1. *System Model and Problem Formulation*

Similar to the previous section, suppose the AUV moves around on the predefined trajectories and periodically broadcasts beacon signals. We assume that it moves on an arbitrary direction in constant velocity for K broadcast periods. Then, it alters the direction and repeats this process. On an arbitrary line, suppose the 3D velocity of the AUV is $\mathbf{v}$. The position of the AUV at the k-th time slot is $\mathbf{x}_k$, the corresponding Doppler estimate is $f_D^{(k)}$, and the position of the target device is $\mathbf{x}$. Then, we can obtain

$$f_k(\boldsymbol{\theta}) = \frac{(\mathbf{x} - \mathbf{x}_k)^T \mathbf{v}}{\|\mathbf{x}^T - \mathbf{x}_k\|} \cdot \frac{f_c}{c} - f_D^{(k)} \approx 0, \qquad (8.18)$$

where c denotes the average underwater sound speed and $\boldsymbol{\theta} = [\mathbf{x}^T, c]^T$ contains the 3D coordinate of the target and the underwater acoustic velocity. Suppose we repeat this process on M lines and we will get MK equations. By stacking all the equations together, we have $\mathbf{f}(\boldsymbol{\theta}) = [f_1(\boldsymbol{\theta}), f_2(\boldsymbol{\theta}), \cdots, f_K(\boldsymbol{\theta})]^T \approx \mathbf{0}$. The maximum likelihood estimate of the parameters is given as

$$\hat{\boldsymbol{\theta}}_{ml} = \arg\min_{\boldsymbol{\theta}} \|\mathbf{f}(\boldsymbol{\theta})\|. \tag{8.19}$$

By solving this optimization problem, we can get an estimate of the target's location. However, this is not a convex problem. Generally, we can employ iterative algorithms to solve them. However, the initial estimate should be carefully chosen to make sure the iterative algorithm converges. To avoid the initial estimation process, a low-complexity two-phase algorithm can be employed. Before introducing the algorithm, we will first talk about the Doppler shift estimation problem.

8.3.2. *Doppler Shift Estimation*

In this section, we will try to answer the following question: how to estimate Doppler-shift, and how is the estimation error distributed? To make the problem mathematically tractable, we will start with the single-path scenario. By removing the multi-path effect, we can better evaluate the impacts of different parameters on system performance. Besides, the theoretical results can still serve as a benchmark. Intuitively, the system performance will degrade in the presence of the multi-path effect. Moreover, the multi-path components play a negligible role in Doppler estimation accuracy, when the sampled sequence is very long [24].

The on-board transmitter of the AUV broadcasts a sinusoidal wave at a frequency of f_c (in Hz). The receiver samples the received signal at f_s (in Hz), and the sampled sequence $\mathbf{s}$ at the target side will be

$$\mathbf{s}[n] = A \sin\left(2\pi(f_c + f_d)/f_s n + \theta\right) + \mathbf{n}_s[n], \tag{8.20}$$

where f_d is the Doppler shift, A is the amplitude of the received signal, and $\mathbf{n}_s$ contains zero-mean Gaussian noise, with a variance of σ^2. Let $\omega = 2\pi(f_c + f_d)/f_s$, and we have

$$\mathbf{s}[n] = A \sin\left(\omega n + \theta\right) + \mathbf{n}_s[n]. \tag{8.21}$$

For a sample number of N, we can obtain the discrete spectrum of $\mathbf{s}$ as $\mathbf{s}_\omega = \text{DFT}\{\mathbf{s}\}$. For a sinusoid wave, the envelope of the spectrum is very close to a scaled sinc function, and there are always two samples in the main lobe if we conduct DFT, as shown in Fig. 8.3. The sampled spectrum

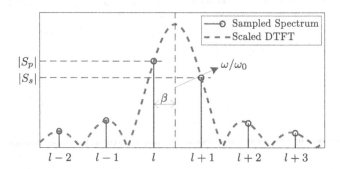

Fig. 8.3.   When $0 \le \beta < 0.5$, the peak appears on index l, while the sub-peak is on $l+1$

is obtained through N-point DFT. Define $\omega_0 = 2\pi/N$, and there must exist $l \in \{0, 1, \cdots, N-1\}$ and $\beta \in [0, 1)$ that satisfy $\omega = (l + \beta)\omega_0$.

Let S_p and S_s represent the two samples in the main lobe. Generally, they have the largest amplitudes in the DFT sequence. When $0 \le \beta < 0.5$, we have $S_p = \mathbf{s}_\omega[l]$ and $S_s = \mathbf{s}_\omega[l+1]$. As has been shown in [25], an estimate β can be obtained as

$$\hat{\beta} = \frac{|S_s|}{|S_s| + |S_p|}. \tag{8.22}$$

It should be noted that both S_s and S_p follow Gaussian distribution, and they have the same variance. In this case, $\hat{\beta}$ can be approximated by Gaussian distribution [25], and the variance is

$$\mathrm{var}\{\hat{\beta}\} \approx \frac{2\sigma^2 r_0(\beta)}{N A^2}, \tag{8.23}$$

where $r_0(\beta)$ is a function of β.

On the other hand, for $\beta \in [0.5, 1)$, β can be estimated as

$$\hat{\beta} = \frac{|S_p|}{|S_p| + |S_s|}, \tag{8.24}$$

and we can prove that the variance of $\hat{\beta}$ is identical to the results in Eq. (8.23).

When β is very close to 0 or 1, this algorithm is not reliable. To be specific, there is a high probability that the sub-peak cannot be correctly identified because it is very weak. This scenario is illustrated in Fig. 8.4. Originally, the peak sample and the sub-peak sample are indexed by l and $l_p = l + 1$. However, β is very close to one, and the sub-peak is very weak,

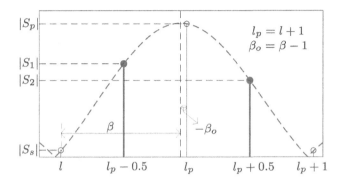

Fig. 8.4.   Refinement of Doppler shift estimation

comparable to the other samples outside the main lobe. In the presence of noise, l cannot be correctly identified.

In this case, an algorithm is proposed in [25] to refine the result. Let l_p be the index of the peak value in the spectrum, and there must exist $\beta_o \in [-0.5, 0.5)$ that satisfies $\omega = (l_p + \beta_o)\omega_0$. We have $l_p + \beta_o = l + \beta$, or $\beta_o = \beta - \lfloor \beta \rceil$ equivalently. As shown in Fig. 8.4, we can take another two samples equally spaced around the peak sample, (i.e., S_1 and S_2), and use them to accurately estimate β_o. The details can be found in [25].

With $\hat{\beta}$, we have the estimate of f_d as

$$\hat{f}_d = f_s(l + \hat{\beta})/N - f_c, \tag{8.25}$$

and the variance is

$$\mathrm{var}\{\hat{f}_d\} \approx \frac{2\sigma^2 f_s^2}{N^3 A^2} r(\beta), \tag{8.26}$$

as is verified by the simulation results in Fig. 8.5.

8.3.3. *Doppler-Based Localization*

Now we have estimated the Doppler shifts and obtained some equations with respect to the target's location and the average underwater sound velocity. In this section, we present a low-complexity two-phase algorithm to solve these equations. In the first phase, the coarse result is obtained by extracting linear constraints on the unknowns. By doing this, we can obtain the localization result with linear algorithms, but we are not fully utilizing the information. Therefore, we add a second phase, during which the localization result will be refined.

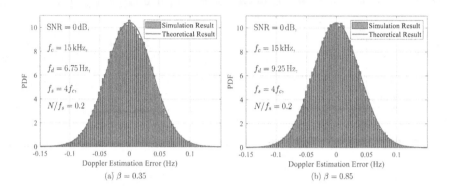

Fig. 8.5.   Numerical and theoretical Doppler estimation error

8.3.3.1.   *Phase I: coarse localization*

As we have discussed, we want to extract linear constraints on the target's
location from the nonlinear equations. The intuition is presented in Fig. 8.6.
As we can see, the AUV moves on a straight line at a constant velocity. At
t_1, it broadcasts the beacon signal, and the target can decide that it is on
a conical surface by estimating the Doppler shift. At t_2, a second conical
surface can be identified. The target must lie at the intersection of these
two conical surfaces, which is a circle, and that circle must lie on a specific
plane. As long as we can find that plane, we can establish a linear equation

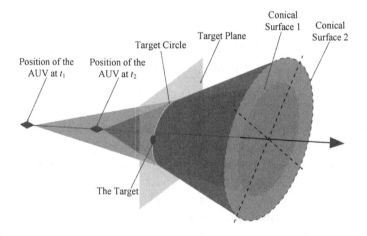

Fig. 8.6.   Geometrical intuition of low-complexity algorithm

with respect to the target's location. When the AUV moves in different directions, we can obtain a series of linear equations, and by solving them, the target's position can be estimated.

Mathematically, suppose the AUV is moving on a straight path at a constant speed $\mathbf{v}$. At the first and the second broadcast periods, the Doppler shift measurements are $f_D^{(1)}$ and $f_D^{(2)}$, respectively. Given that the positions of the AUV at these two periods are $\mathbf{x}_1$ and $\mathbf{x}_2$, the following equations can then be constructed

$$(\mathbf{x}^T - \mathbf{x}_1^T)\mathbf{v} = c_1 d_1, \tag{8.27a}$$

$$(\mathbf{x}^T - \mathbf{x}_2^T)\mathbf{v} = c_2 d_2, \tag{8.27b}$$

where $c_i = c f_D^{(i)}/f_c$ and $d_i = \|\mathbf{x}^T - \mathbf{x}_2\|$ ($i = 1, 2$). As shown in [12], by defining $w = \mathbf{x}^T \mathbf{v}$, we can transform Eq. (8.27a) and Eq. (8.27b) into a quadratic equation:

$$aw^2 + bw + q = 0, \tag{8.28}$$

where a, b, q are dependent on c_i, $f_D^{(i)}$, $\mathbf{x}_i$ and $\mathbf{v}$.

By solving Eq. (8.28), we will obtain two roots as

$$\hat{w} = \frac{-b \pm \sqrt{b^2 - 4aq}}{2a}. \tag{8.29}$$

These two roots represent two parallel planes, and the target can only lie on one of them. In Fig. 8.7, we show these two planes in top view. As we can see, the intersection of two conical surfaces lies on the true target plane. However, if we extend the second conical surface to the opposite

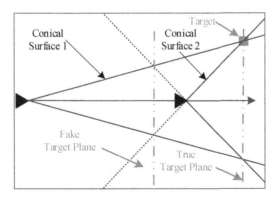

Fig. 8.7.   Fake target surface and true target surface

direction, it will have another intersection with the first conical surface, and that gives us the fake solution. To identify the true target plane, we notice that ω must satisfy the following inequality:

$$(\omega - \mathbf{x}_i^T \mathbf{v}) f_D^{(i)} = (\omega - \mathbf{x}_i^T \mathbf{v})^2 f_c / d_i / c \geq 0, \qquad (8.30)$$

where we implicitly replace $f_D^{(i)}$ with the left-hand side of Eq. (8.18).

Note that we can get multiple measurements on an arbitrary line and combine them to get a better estimate of ω and average out noise. Assume we have obtained the valid solution as $\hat{\omega}$, and the target surface will be determined by

$$\mathbf{v}^T \mathbf{x} = \hat{\omega}. \qquad (8.31)$$

This process will be repeated in M directions, and we will get M linear equations. We can then employ the LS estimator to obtain the coarse estimate of the target's location.

As we can see, the complexity of this algorithm grows linearly with the number of Doppler measurements. Besides, it should be noted that c is assumed to be known in this algorithm. Generally, it is not far away from $1500\,\mathrm{m/s}$, and we can employ this value for the coarse estimation.

8.3.3.2. *Phase II: refinement of the result*

In *Phase I*, we are only partially using the information provided by the Doppler measurements. For example, in Fig. 8.6, the Doppler estimates tell us that the target lies on a specific circle, but we are extending the searching area to a plane, which leads to information loss. As a result, we should find a way to further extract the available information and try to improve localization accuracy.

The coarse estimate of θ is $\hat{\theta}_c = [\hat{\mathbf{x}}_c^T, \hat{c}_c]^T$, where $\hat{\mathbf{x}}_c$ is the coarse position of the target obtained from Phase I, and $\hat{c} = 1500$ m/s. Let $\Delta\theta$ be the estimation error, i.e., $\hat{\theta}_c = \theta + \Delta\theta$, and $\Delta\theta$ can be approximated by the first order Taylor expansion as

$$\mathbf{f}(\hat{\theta}_c) - \mathbf{f}(\theta) \approx \mathbf{H}_c \Delta\theta. \qquad (8.32)$$

$\mathbf{H}_c$ is the Jacobian matrix evaluated at $\hat{\theta}_c$.

$\Delta\theta$ can then be estimated with the weighted LS algorithm:

$$\Delta\hat{\theta} \approx (\mathbf{H}_c^T \mathbf{W} \mathbf{H}_c)^{-1} \mathbf{H}_c^T \mathbf{W} \mathbf{f}(\hat{\theta}_c), \qquad (8.33)$$

where $\mathbf{W}$ is the weight matrix. From the previous discussions, we know that the Doppler estimation error is different for different time slots. The

Doppler measurements with smaller estimation error should be given larger weights during the numerical computation. The optimal choice of $\mathbf{W}$ is discussed in [25].

Then, the coarse estimate $\hat{\boldsymbol{\theta}}_c$ can be refined as

$$\hat{\boldsymbol{\theta}} = \hat{\boldsymbol{\theta}}_c - \Delta\hat{\boldsymbol{\theta}}. \tag{8.34}$$

It should be noted that this process can be iterated to further improve the estimation accuracy. Based on the simulation results, one or two iterations should be enough to provide very accurate localization results.

8.3.3.3. *CRLB analyses*

As we have demonstrated in Sec. 8.3.2, the Doppler estimation error can be well approximated by zero-mean Gaussian distribution, and the variance is given by Eq. (8.26). Based on this conclusion, we can derive the FIM, which can be used to quantify the amount of the target's location information that can be extracted from the Doppler estimates.

When N is very large, the asymptotic FIM is given in [25] as

$$\mathbf{F} \sim \mathbf{H}^T \boldsymbol{\Sigma}^{-1} \mathbf{H}, \tag{8.35}$$

where $\boldsymbol{\Sigma}$ is the covariance matrix of the Doppler shift estimates and $\mathbf{H}$ is the Jacobian matrix. Because the elements in $\boldsymbol{\Sigma}$ are proportional to the cube of the sequence length, i.e., N, the variance of positioning error also decreases cubically with N.

8.3.4. *Numerical Evaluations*

In this section, we will conduct simulations to verify the analytical results in previous sections. We will investigate how different parameters contribute to the overall performance.

8.3.4.1. *Impact of iteration number and SNR*

As we have briefly mentioned, if we repeat the refinement process in *Phase II*, the localization accuracy can be improved. Simulations are conducted and the results are presented in Fig. 8.8. f_c is set as 15 kHz, the sampling time is fixed as 0.1 s, and the sampling frequency is 60 kHz. When the iteration number (N_{iter}) equals zero, there is a discernible gap between the CRLB and the MSE of the proposed method, especially in the low SNR regime. However, by adding one iteration, the localization accuracy

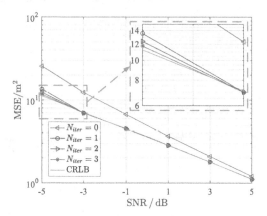

Fig. 8.8.   Impact of iteration number and SNR on localization accuracy

can be substantially improved. By increasing the iteration number to 2 or 3, the localization error will decrease continuously, but the performance gain is negligible. Besides, we notice that as SNR increases, localization error decreases constantly, because Doppler shift can be more accurately estimated. Also, we can see that the variance of 3D localization error is inversely proportional to the SNR.

8.3.4.2.   Impact of N and SNR

Intuitively, as we increase N or SNR, the accuracy of Doppler estimation will be improved. As a result, localization accuracy should be improved. In this part, we will evaluate the system performance for different N and SNR values. The sampling frequency is fixed. Therefore, the increase of N is equivalent to the increase of sampling time. Based on our previous discussions, the variance of positioning results is inversely proportional to N^3. The simulation results are presented in Fig. 8.9. Every time we double the number of samples, the MSE of localization result will be reduced by a factor of 8, which is consistent with the previous analysis. Besides, we notice that for $N = 0.05f_s$, the proposed algorithm has a significant performance gap compared to the CRLB. This is because the analysis is based on the first-order Taylor expansion, which is only accurate for small Doppler shift estimation error. When N is not large enough, Doppler estimation error will be significant and the theoretical result is no longer accurate.

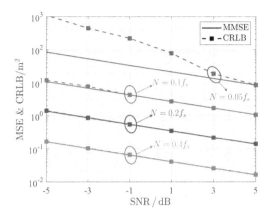

Fig. 8.9.   Impact of N on localization accuracy

8.3.4.3. *Impact of AUV's trajectory length*

Suppose the AUV moves on M directions and broadcasts the beacon signals for K time periods on each of them. When we increase M and K, the localization error is expected to decrease. In this section, we conduct simulations for different combinations of M and K. The results are shown in Fig. 8.10. As we can see, M is increased from 2 to 4, and then to 6, while K varies in $\{4, 6, 8\}$. Generally, by increasing M and K, more measurements can be obtained, and better localization accuracy can be achieved. However, the impact is more complicated compared to that of N. As has

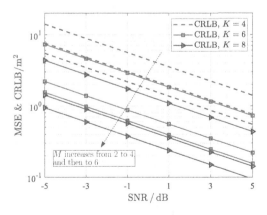

Fig. 8.10.   Impact of M and K on localization accuracy

been pointed out by many researchers, the trajectory of the AUV has a significant impact on localization accuracy. Therefore, depending on the trajectory of the AUV, the increase of M and K may show a significant or negligible influence on performance.

8.4. Joint Estimation of Position and Velocity Based on LFM Signals

In Sec. 8.2, a single AUV is employed to locate a target equipped with a hydrophone. In Sec. 8.3, this requirement is relaxed, and the target only needs an acoustic device being able to generate or receive sinusoidal waves at a fixed frequency. However, what if the target is totally silent and cannot transmit or receive any acoustic signals? These kinds of examples include icebergs, whales, leaked oil, etc. For these scenarios, we can employ the synchronized and localized nodes as anchors for proactive target detection and tracking, which will be discussed in this section.

8.4.1. *Background*

Conventionally, short pulse signals with low duty cycles are used as the probe signals. The operator only has one detection opportunity for every cycle, which is generally designed to be very long to detect objects far away from the network. The use of short pulses was inevitable in the past because the sonar systems had limited dynamic range, which forced the transmit signal to have a steady envelope. However, the state-of-the-art sonar systems have much larger dynamic ranges, and the duty cycle is no longer severely limited, which makes it possible to implement the Continuous Active Sonar (CAS).

In CAS, probe signals with very high duty cycles will be used for target detection. Compared with the conventional PAS (Pulsed Active Sonar), CAS has the following advantages:

(1) CAS has higher detection probabilities. For underwater objects, especially large-sized objects, glint noise is a big problem. Because of glint noise, the strength of the received reflection varies with time and is mostly weak. As a result, the pulsed signals have a large miss rate.
(2) CAS suppresses false alarm rate. In shallow waters, there are many unstable reflections. The PAS cannot filter these components, while CAS can average them out through time diversity.

(3) CAS can improve tracking performance. Due to the low duty cycle, PAS cannot provide continuous information with respect to the target, which leads to target ambiguity.
(4) CAS works at much lower power, which is environmentally friendly. The negative impact of sonar on underwater animals can be minimized because the signal strength is at the ambient noise level.

In [26], the authors employed CAS for target localization by jointly estimating the target's distance and DoA. To be specific, the localization result is the coordinate that maximizes the likelihood function of the ToA and DoA measurements. In [27], joint estimation of the target's velocity and the position was considered. In [28], an experiment was conducted to show that the CAS can achieve much better performance at lower SNR, compared with the conventional systems.

For CAS, the Linear Frequency Modulated (LFM) signals are preferred because they achieve a great balance between time and frequency domain resolutions, allowing the simultaneous estimation of both target distance and radial velocity. LFM signals have time-variant spectrums, and we cannot use the conventional Fourier transform to analyze the spectrum of the received signals. Instead, Wigner distribution should be computed to accurately estimate the initial frequency and the frequency rate of the received signal. However, this involves very high computational complexity, leading to the difficulty of real-time signal processing. In [29], LFM signals are used for probing, and a band of lag-Doppler filters is used for joint estimation of delay and Doppler shift. The filters are designed based on uniform sampling in velocity and distance. However, it is not self-adaptive. Fortunately, the surge of the Fractional Fourier Transform (FrFT) provides an alternative [30, 31]. To be specific, it is shown that the FrFT of a given signal corresponds to a rotation in the Wigner distribution [30].

In an acoustic sensor network, some of the nodes will periodically broadcast LFM signals, which will hit the targets, get reflected, and is received by the other nodes. Depending on the target's position and velocity, the received signals will also be LFM signals of different frequencies and frequency rates. The receiving node will then conduct FrFT on the received signal, and we can get the 2-dimensional (2D) spectrum of the received signal. For an LFM signal, we can always find a peak in the spectrum, and its position is dependent on the initial frequency and frequency rate. Therefore, we can estimate the target's distance and radial velocity with respect to the receiving node. With enough receiving nodes, we can then estimate the target's location and velocity.

A major challenge of this system is the computational complexity. Although the fast FrFT algorithms proposed in [31] have already reduced the computational complexity to a great extent, we still need to process a very large 2D spectrum. If there are multiple objects, it will become highly complicated. To solve this problem, we can employ machine learning techniques to first roughly estimate the position of the peak, and then conduct over-sampling on the small area around the peak. This is possible depending on the following observation: if a target exists, we will be able to find an "X" pattern on the spectrum, and the cross-point is dependent on the target's location and velocity [32]. If multiple targets exist, there will be multiple "X" patterns.

For the detection of the "X" patterns, the Convolutional Neural Networks (CNNs) can be used, because they are specifically designed for pattern recognition. The major advantage of this approach is that it allows us to compute the discrete spectrum with a much larger sampling interval. With the under-sampled spectrum, although we won't be able to accurately estimate the position of the peak, the "X" patterns can still be preserved, which allows us to perform the coarse estimation. The computational complexity can thus be significantly reduced.

8.4.2. *System Model and the Discrete FrFT Algorithm*

We will start with a simple 2D model with only two nodes. Node 0 is transmitting a probe signal periodically, while node 1 is listening. A target is moving in the surveillance area. The signal transmitted from node 0 is reflected by the target and received by node 1, as illustrated in Fig. 8.11. Suppose node 0 and node 1 are located at $\mathbf{x}_0$ and $\mathbf{x}_1$, while the target moves

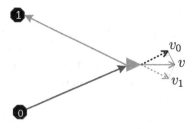

Fig. 8.11.   System model illustration

at a constant speed of $\mathbf{v}$. Node 0 transmits a LFM signal from $t = 0$ to $t = T$, given by

$$s(t) = Ae^{j(2\pi f_0 t + k\pi t^2)}, \; t \in [0, \; T], \tag{8.36}$$

where A is the amplitude.

The received signal at node 1 will be

$$r(t) = P_0 s(\rho(t - \tau)), \tag{8.37}$$

which is another LFM signal. In Eq. (8.37), the received signal has three important parameters: P_0, τ, and ρ. P_0 is the propagation loss; τ is the delay of the received signal, and it is proportional to the sum of the distances between nodes and the target, at the time the signal first reached the target. ρ is the time scaling factor (Doppler rate), given by

$$\rho = \frac{c - v_0}{c + v_1}, \tag{8.38}$$

where v_0 and v_1 are the radial speeds of the target with respect to node 0 and 1, and c is the average underwater sound speed. If the target is moving away from the node, the velocity is positive. Otherwise, it will be negative. Assume the target's position is $\mathbf{x}$ at $t = 0$, we have

$$\begin{aligned} v_0 &= \frac{(\mathbf{x} - \mathbf{x}_0)^T \mathbf{v}}{\|\mathbf{x} - \mathbf{x}_0\|}, \\ v_1 &= \frac{(\mathbf{x} - \mathbf{x}_1)^T \mathbf{v}}{\|\mathbf{x} - \mathbf{x}_1\|}. \end{aligned} \tag{8.39}$$

Then, the signal will first reach the target roughly at $t = \tau_0 = \|\mathbf{x} - \mathbf{x}_0\|/(c - v_0)^*$, when the target is located at $\mathbf{x}_{\tau_0} = \mathbf{x} + \mathbf{v}\tau_0$. As a result, we have

$$\|\mathbf{x}_0 - \mathbf{x}_{\tau_0}\| + \|\mathbf{x}_1 - \mathbf{x}_{\tau_0}\| = c\tau. \tag{8.40}$$

Assume the maximum scanning distance is d_{max}, and the maximum delay will be $\tau_{max} = 2d_{max}/c$. After frequency mixing at the receiver, the received signal will go through a low-pass filter, and the result will be

$$\begin{aligned} r(t) &= A_0 e^{j[(2\pi f_0(1-\rho)t + k\pi(1-\rho^2)t^2 + 2k\pi\rho^2\tau t)]} \\ &= A_0 e^{j(2\pi \tilde{f}_0 t + \pi \tilde{k} t^2 + \theta)}, \end{aligned} \tag{8.41}$$

where A_0 is the amplitude of the signal after filtering. $\tilde{f}_0$ and $\tilde{k}$ are given as

$$\tilde{f}_0 = f_0(1 - \rho) + k\rho^2\tau \text{ and } \tilde{k} = k(1 - \rho^2), \tag{8.42}$$

---

*Here we are assuming that the v_0 is constant from $t = 0$ to $t = \tau_0$.

for $t \in [\tau_{max}, T]$, and θ is the phase delay, and it is of no significance for the localization purpose.

Apparently, this is a new LFM signal, whose initial frequency and frequency rate are given as $\tilde{f}_0$ and $\tilde{k}$ in Eq. (8.42). By estimating $\tilde{f}_0$ and $\tilde{k}$, we can compute the values of ρ and τ, which are directly dependent on the Doppler shift and delay. These estimates will then eventually lead to equations concerning the target's distance and radial velocity. For a large underwater sensor network, by incorporating the information from multiple nodes, we can estimate the target's location and velocity.

The major challenge here is the accurate estimation of the parameters of the LFM signals at the receiving nodes. The FrFT has been well developed as a powerful tool for this purpose. The a-th order FrFT of $r(t)$ is given as

$$R_a(u) = \int_{-\infty}^{\infty} K_a(u,t) r(t) dt. \tag{8.43}$$

$K_a(u,t)$ is the kernel function given as

$$K_a(u,t) = \sqrt{1 - j \cot \phi} \, e^{j\pi(u^2 \cot \phi - 2 \csc \phi ut + t^2 \cot \phi)}, \tag{8.44}$$

where $\phi = a\pi/2$. For $a = 0$ or $a = \pm 2$, the kernel approaches $K_0(u,t) = \delta(u - t)$ and $K_{\pm 2}(u,t) = \delta(u + t)$, respectively. $a = 1$ gives us the conventional Fourier transform. The FrFT has two important characteristics:

$$R_a(u) = R_{a+4}(u) \text{ and } R_a(u) = R_{a+2}(-u). \tag{8.45}$$

Therefore, we only need to conduct the FrFT for $a \in [-1, 1]$. As we can see, the FrFT tries to decompose the received signal into a series of LFM signals, and we will obtain a 2D spectrum. In the spectrum, assume we have found the peak at $[\hat{a}, \hat{u}]$:

$$[\hat{a}, \hat{u}] = \arg \max_{a,u} |R_a(u)|. \tag{8.46}$$

Obviously, the absolute value of $R_a(u)$ will be maximized for

$$f_0 = u \csc \phi \text{ and } k = -\cot \phi. \tag{8.47}$$

Therefore, the initial frequency and frequency rate can be estimated as

$$\hat{f}_0 = \hat{u} \csc \hat{\phi} \text{ and } \hat{k} = -\cot \hat{\phi}, \tag{8.48}$$

where $\hat{\phi} = \hat{a}\pi/2$.

The direct computation of the FrFT introduces high complexity. In [31], a fast discrete algorithm was proposed. An example of the implementation of this algorithm is shown in Fig. 8.12. The initial frequency is 120 Hz,

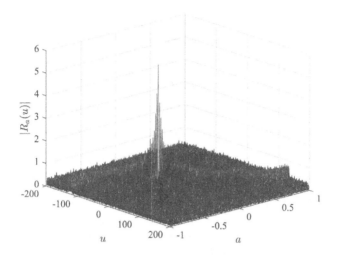

Fig. 8.12. An example of the fast discrete FrFT

and the frequency rate is 4.2, with a sampling rate of 400 Hz. Suppose the signal lasts for 4 seconds, the amplitude of $R_a(u)$ is given in Fig. 8.12. The peak is roughly located at $a = -0.149$ and $u = -27.795$.

8.4.3. *Position and Velocity Estimation of the Targets*

Suppose $M + 1$ nodes are distributed in the area of interest, indexed from 0 to M. Node 0 is periodically broadcasting a LFM signal, while the other nodes are all listening. Assume the target moves at constant speed $\mathbf{v}$. The radial velocity with respect to the m-th receiver will be

$$v_m = \frac{(\mathbf{x} - \mathbf{x}_m)^T \mathbf{v}}{\|\mathbf{x} - \mathbf{x}_m\|}, \tag{8.49}$$

where $\mathbf{x}_m$ and $\mathbf{x}$ are the positions of the m-th node and the target at $t = 0$.

8.4.3.1. *Coarse localization*

Assume the reflected signal is received at the m-th node. The initial frequency f_m and frequency rate k_m are given as

$$\begin{aligned} f_m &= f_0(1 - \rho_m) + k\rho_m^2\tau_m, \\ k_m &= k(1 - \rho_m^2). \end{aligned} \tag{8.50}$$

Suppose f_m and k_m are estimated as $\hat{f}_m$ and $\hat{k}_m$ based on the fast FrFT algorithm. The estimates of τ_m and ρ_m can be computed as

$$\hat{\rho}_m = \sqrt{1 - \hat{k}_m/k},$$
$$\hat{\tau}_m = \frac{\hat{f}_m - f_0(1 - \hat{\rho}_m)}{k\hat{\rho}_m^2}. \tag{8.51}$$

ρ_m and τ_m are the time scaling factor and the delay for the m-th node.

Based on $\hat{\tau}_m$ and $\hat{\rho}_m$, we can obtain two equations

$$\hat{\tau}_m = \|\mathbf{x}_{\tau_0} - \mathbf{x}_m\|/c + \|\mathbf{x}_{\tau_0} - \mathbf{x}_0\|/c,$$
$$\hat{\rho}_m = \frac{c - v_0}{c + v_m}. \tag{8.52}$$

With M listening nodes, $2M$ equations will be available, and five unknowns are involved: the 2D coordinate and velocity, and the average underwater sound speed. These nonlinear equations are very complicated and not easy to solve. Therefore, we will simplify the equations to roughly estimate the unknowns first, as we did in the previous two sections.

Define $f_m(\mathbf{x}_{\tau_0}, c)$ as

$$f_m(\mathbf{x}_{\tau_0}, c) = \|\mathbf{x}_{\tau_0} - \mathbf{x}_m\| + \|\mathbf{x}_{\tau_0} - \mathbf{x}_0\| - c\hat{\tau}_m, \tag{8.53}$$

and the LS estimator is

$$[\hat{\mathbf{x}}_{\tau_0}, \hat{c}] = \arg\min_{\mathbf{x}_{\tau_0}, c} \sum_{m=1}^{M} |f_m(\mathbf{x}_{\tau_0}, c)|^2. \tag{8.54}$$

The objective function is not convex, and we need to conduct coarse estimation first to get an approximation of the optimal position before the iterative algorithms can be employed to refine the result.

The first step is to rewrite Eq. (8.53) as

$$\|\mathbf{x}_{\tau_0} - \mathbf{x}_m\| = -\|\mathbf{x}_{\tau_0} - \mathbf{x}_0\| + c\hat{\tau}_m. \tag{8.55}$$

By doubling both sides, we have

$$\|\mathbf{x}_{\tau_0}\|^2 + \|\mathbf{x}_m\|^2 - 2\mathbf{x}_m^T\mathbf{x}_{\tau_0} = d_0^2 - 2c\hat{\tau}_m d_0 + c^2\hat{\tau}_m^2, \tag{8.56}$$

where $d_0 = \|\mathbf{x}_{\tau_0} - \mathbf{x}_0\|$. Then we minus the m-th equation from the first one and we will obtain

$$\|\mathbf{x}_m\|^2 - \|\mathbf{x}_1\|^2 + 2\mathbf{x}_1^T\mathbf{x}_{\tau_0} - 2\mathbf{x}_m^T\mathbf{x}_{\tau_0}$$
$$= 2(\hat{\tau}_1 - \hat{\tau}_m)cd_0 + (\hat{\tau}_m^2 - \hat{\tau}_1^2)c^2. \tag{8.57}$$

With these linear equations, the LS estimate of the target's position can be obtained as $\hat{\mathbf{x}}_{\tau_0,c}$, while the average underwater acoustic speed can be estimated as $\hat{c}_c$. These results are just coarse estimates, but they should be reasonably close to the true value. Thus, we can use them as the initial estimate and employ iterative algorithms to refine the result.

8.4.3.2. *Refined location and velocity estimation*

Define $\boldsymbol{\theta} = [\mathbf{x}_{\tau_0}^T, c]^T$. By stacking all the equations together as $\mathbf{f}(\boldsymbol{\theta})$, the maximum likelihood estimate of the parameters can be obtained through the Newton method. Based on the coarse localization result, we can construct $\hat{\boldsymbol{\theta}}_0 = [\hat{\mathbf{x}}_{\tau_0,c}^T, \hat{c}_c]^T$ as the initial estimate of $\boldsymbol{\theta}$.

Suppose $\boldsymbol{\theta}$ is estimated as $\hat{\boldsymbol{\theta}}_k$ in the k-th iteration, we can update the estimate in the $(k+1)$-th iteration as

$$\hat{\boldsymbol{\theta}}_{k+1} = \hat{\boldsymbol{\theta}}_k - (\mathbf{F}_k^T \mathbf{F}_k)^{-1} \mathbf{F}_k \mathbf{f}(\hat{\boldsymbol{\theta}}_k), \tag{8.58}$$

where $\mathbf{F}_k$ is the Jacobian matrix evaluated at $\boldsymbol{\theta} = \hat{\boldsymbol{\theta}}_k$.

Generally, one or two iterations should lead to convergence, which means low computational complexity. Suppose the refined localization result is $\hat{\mathbf{x}}_{\tau_0}$ and the underwater acoustic velocity is estimated as $\hat{c}$. The next step is to estimate the target's velocity.

From the second part of Eq. (8.52), we have

$$c - \frac{\mathbf{x}^T - \mathbf{x}_0^T}{\|\mathbf{x} - \mathbf{x}_m\|}\mathbf{v} = \rho_m c + \rho_m \frac{\mathbf{x}^T - \mathbf{x}_m^T}{\|\mathbf{x}^T - \mathbf{x}_m^T\|}\mathbf{v}. \tag{8.59}$$

Because we are assuming that the target's radial velocity is constant during the sampling time, we can replace $\mathbf{x}$ with $\mathbf{x}_{\tau_0}$ in Eq. (8.59). The incurred error will be negligible. To be specific, we can obtain the following approximation

$$\left(\frac{\rho_m(\hat{\mathbf{x}}_{\tau_0} - \mathbf{x}_m)}{\|\hat{\mathbf{x}}_{\tau_0} - \mathbf{x}_m\|} + \frac{(\hat{\mathbf{x}}_{\tau_0} - \mathbf{x}_0)}{\|\hat{\mathbf{x}}_{\tau_0} - \mathbf{x}_0\|}\right)^T \mathbf{v} \approx (1 - \hat{\rho}_m)\hat{c}, \tag{8.60}$$

which is a linear equation of the target's velocity. There are M equations like this and they can be easily solved with the LS method.

8.4.4. *CNN-Based Target Detection*

For the discussed underwater localization system, the proactive nodes are periodically broadcasting the LFM signals, with a period of T, of several seconds. To achieve real-time scanning, the computation load must be handled very fast. However, the computation of Fig. 8.12 is still very challenging, even when we use the fast algorithm. Another problem is that the accurate localization of the peak is highly dependent on the sampling interval. With smaller intervals, higher accuracy can be obtained, at the price of increased computational cost.

Apart from the computational cost, how to identify the existence of targets is also a huge problem. An intuitive idea is to set up a threshold and claim the existence of targets whenever there are samples larger than the threshold. However, the spectrum is not sparse, and we can see quite a few side peaks comparable to the highest peak, even when we only have one target in the surveillance area. Also, if multiple objects exist, the situation will become more complicated.

These two problems can be simultaneously solved by incorporating a CNN in the system. With this CNN, we can first compute the coarse spectrum with a low sampling rate, based on which, we will detect the existence of the targets. Besides, CNN can tell us the rough locations of these peaks in the spectrum. Then, around each peak, over-sampling can be conducted to refine the result.

Practically, we cut a small portion of the map in Fig. 8.12, and normalize all the values to [0, 255]. We can then draw the top view of that partial map in grayscale, as shown in Fig. 8.13. An interesting observation is that we can always find an "X" pattern around the peak value. The position, orientation, and width of the pattern varies with the parameters, including initial frequency, frequency rate, signal duration, sampling frequency, and so on. Even when we sample the spectrum at a much lower rate, this pattern can still be preserved. As a result, an intuitive idea is to search this pattern in an under-sampled spectrum and then conduct a fine search around the target area to improve the accuracy. By doing this, high accuracy can be obtained without high complexity. The fundamental reason for the performance improvement is that the "X" pattern can be very well

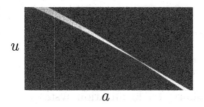

Fig. 8.13.   Top view of spectrum in Fig. 8.12 around the peak

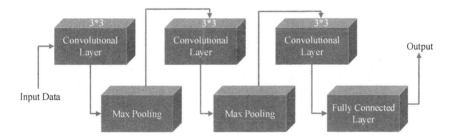

Fig. 8.14. Structure of the CNN, including three convolutional layers, two pooling layers, and one fully connected layer

preserved even when we conduct under-sampling on the 2D spectrum. As a result, the powerful CNN allows us to find the rough location of the peak based on a low-resolution 2D spectrum.

The structure of the employed CNN is shown in Fig. 8.14. The input is the 99 × 99 figures obtained from the spectrum, while the output is "Negative" or "Positive", depending on whether the targets are identified in the figures. There are in total six layers: three convolutional layers, two max-pooling layers, and one fully connected layer. Each pixel stores a number varying from 0 to 255. Every figure will first go into a convolutional layer, with eight filters of size 3 × 3. The convolutional layer will be followed by a batch normalization layer and a ReLU layer. A max-pooling layer is added to eliminate redundancy.

The second convolutional layer has 16 filters of size 3 × 3, while the third convolutional layer has 32 filters of the same size. The increase of the filters can help to extract the high-level structures. Lastly, we have a fully connected layer with ten outputs, followed by the softmax layer and the classification layer.

Because the input has a dimension of 99×99, while the original spectrum is very large, we need to first conduct under-sampling on the 2D spectrum. The default sampling interval is 100, if not specified. Then we can get a much smaller map, and a carpet search will be conducted with the trained CNN. The under-sampled map will be divided into tens of blocks, which will be fed into the CNN sequentially. This structure will apparently allow us to detect multiple objects.

One input figure contains around ten thousand pixels, which is quite large. However, due to the sparse structure of the CNN, only a very small number of parameters are required for training. To be specific, for each convolutional layer, one filter has only 3 × 3 = 9 parameters. With the 56

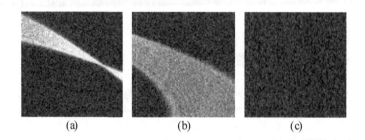

Fig. 8.15. Examples of (a) a positive case, and (b) and (c) negative cases

filters, only hundreds of parameters are needed to describe the whole network. Therefore, we can see that the number of parameters is not dependent on the input size, which means the training complexity grows linearly with the input size. After three convolutional layers, the output is then fed into a fully connected layer, but the size is small and the complexity is low. These advantages all come from the fact that CNN can extract sparse features from the pictures.

Apparently, the major challenge is how to generate and label the training data sets. One choice is to generate data in MATLAB through simulations. The generated 2D spectrum will be divided into small blocks. Codes are written to automatically label the generated blocks for training purposes. In our case, there are only two labels: "Negative" means no target detected, while "Positive" indicates a detected pattern. Examples are given below in Fig. 8.15.

In Fig. 8.15 (a), we have a positive case, where the "X" pattern can be clearly observed. White noise presents out of that pattern and the noise is weaker than the signal. In (b), we have a negative case. Part of the tail of a pattern is captured, and we can infer that we should be able to find the pattern by shifting the window to the top-left. In (c) we have another negative case, and the map contains pure noise.

Based on the collected data, we conducted data augmentation. The basic idea is to shift, rotate, and scale the figures of the positive cases, and obtain variations of the original data. This will help to improve the robustness of the CNN, by training it to recognize the shifted, rotated, and scaled patterns. In practical measurements, the position, width, and orientation of the "X" pattern are dependent on many parameters, including sampling rate, sampling time, and also the under-sampling process. For different parameters, the "X" patterns may look quite different for the same LFM signal. Therefore, data augmentation is an important step to improve the robustness of the CNN.

Simulations for different distance-velocity combinations were conducted. For each combination, the discrete FrFT was utilized to compute the two-dimensional spectrum. Then, the program will automatically divide the spectrum into small blocks of 99×99, and each piece will be stored as a png file. Those figures with the "X" patterns will be stored in the "Positive" folder, while the others will be stored in the "Negative" folder. For the positive cases, data augmentation will be conducted to enlarge the data set.

8.4.5. *Numerical Evaluations*

The initial frequency of the probe signal is chosen as 1 kHz, and the frequency rate is 50 Hz/s. The scanning period is eight seconds, and the maximum scanning distance is three kilometers. As a result, for each period, we will conduct FrFT on the received signals in the last four seconds to avoid ambiguity. The average underwater velocity is 1500 m/s. After frequency mixing of the received signal and the local probe signal, we will get the signal in Eq. (8.42). This signal will then be sampled at the frequency of $f_s = 400$ Hz, and the fast FrFT algorithm can then be employed to obtain the spectrum of the signal.

There are nine anchors distributed on the seafloor, one of which is actively broadcasting the probe signal while the others are silent. Without loss of generality, we build a 2D coordinate system, centered at the proactive anchor, and the silent anchors are located at: $[-1000, -1000]^T$; $[-1000, 1000]^T$; $[1000, -1000]^T$; $[1000, 1000]^T$; $[-2000, -2000]^T$; $[-2000, 2000]^T$; $[2000, -2000]^T$; $[2000, 2000]^T$. Suppose the target is located at $[0, 1500]^T$ at $t = 0$, and it is moving at a velocity of $[2, 1]^T$ m/s.

8.4.5.1. *CNN-Based Target Detection*

For the CNN-based target detection, we conduct the training with 500 and 750 training samples. The trained network is then used for target detection for 300 samples. The results can be found in Table 8.1. In Table 8.1, "Pos." and "Neg." are abbrevations of "Positive" and "Negative", respectively. "True Positive" means that the target exists, and is successfully detected by the network; "True Negative" means that the target does not exist and is correctly reported as "Negative"; "False Positive" means the target

Table 8.1.   Validation statistics

	500 Training Samples		750 Training Samples	
	Pos. (%)	Neg. (%)	Pos. (%)	Neg. (%)
True (%)	99.27	97.27	99.67	99
False (%)	2.73	0.73	1	0.33
Error Rate (%)	1.73		0.67	

does not exist but is reported as "Positive"; "False Negative" means the
"Positive" is reported to be "Negative". By increasing the training samples
from 500 to 750, the overall error rate will decrease from 1.73 to 0.67 %.
For practical application, this error rate is totally acceptable. Because the
network keeps scanning periodically, the probability that a target is not
detected in two consecutive periods is negligible.

8.4.5.2. *Joint position and speed estimation*

For the joint estimation of the target's location and speed, the accuracy
is dependent on various parameters. In this part, we will evaluate how
SNR, anchor number, and iteration number will contribute to the system
performance.

   In Fig. 8.16, the impact of SNR and M on localization accuracy is
evaluated. The SNR varies from -10 to 0 dB. As we increase the SNR,

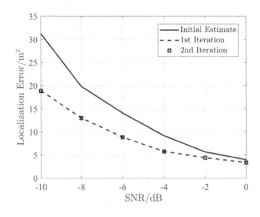

Fig. 8.16.   Localization error for different iteration numbers

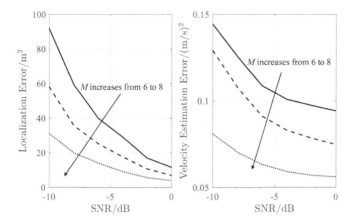

Fig. 8.17. Localization error for different M

the localization error will gradually decrease. Without iteration, we can see that the localization error is at level 2 to 6 meters. If we conduct one iteration, the accuracy will be improved by around one meter at -10 dB. However, for high SNR, the improvement will gradually become negligible. If we further increase the iteration number to 2, we can see that the results are almost identical for those of one iteration. As a result, we can conclude that one iteration is totally enough.

In Fig. 8.17, simulations are conducted for $M \in \{6, 7, 8\}$. Every time we add one extra anchor, the average localization error will decrease by around one meter. This is a well-known effect because the increase of anchor number leads to the decreased GDOP (Geographical Dilution of Precision), which is proportional to the variance of the localization error. For a very large underwater sensor network, M is generally very large, which leads to great positioning accuracy. Practically, the distribution of the anchors also contributes to the localization accuracy and thus should be carefully manipulated.

Based on the localization result, we can now estimate the target's velocity. This can be easily done through Eq. (8.60). The simulations are conducted for different anchor numbers, and the results are presented on the right-hand side of Fig. 8.17. Similar to the localization error, we can see that the increase of anchor number leads to improved performance. The velocity estimation error is at the level of 0.2 to 0.3 m/s, which renders a 10% estimation error.

8.5. Summary

In this chapter, we investigated the smart AUV-assisted underwater localization, time synchronization, and velocity estimation. In the first section, we talked about the ToA-based localization and time synchronization of underwater acoustic sensor networks. In the second section, we showed that the Doppler shift can also be used for target localization and tracking. Compared with the ToA-based approach, the Doppler shift-based method can work on simplified hardware. In the third section, we investigated the localization of silent objects through LFM signals. In all these three scenarios, we presented low-complexity algorithms, because the real-time processing of the collected information is very important. We can see that by using smart AUVs as mobile anchors in underwater acoustic sensor networks, great performance can be achieved at lower costs.

References

1. M. Erol, L. F. M. Vieira, and M. Gerla, "AUV-aided localization for underwater sensor networks," in *International Conference on Wireless Algorithms, Systems and Applications (WASA 2007)*, Chicago, Illinois, USA, Aug. 2007, pp. 44–54.
2. D. M. Crimmins, C. T. Patty, M. A. Beliard, J. Baker, J. C. Jalbert, R. J. Komerska, S. G. Chappell, and D. R. Blidberg, "Long-endurance test results of the solar-powered auv system," in *OCEANS 2006*, Boston, MA, USA, Sept. 2006, pp. 1–5.
3. J. Liu, Z. Wang, J. H. Cui, S. Zhou, and B. Yang, "A joint time synchronization and localization design for mobile underwater sensor networks," *IEEE Transactions on Mobile Computing*, vol. 15, no. 3, pp. 530–543, March 2016.
4. C. R. Berger, S. Zhou, P. Willett, and L. Liu, "Stratification effect compensation for improved underwater acoustic ranging," *IEEE Transactions on Signal Processing*, vol. 56, no. 8, pp. 3779–3783, Aug. 2008.
5. R. Diamant and L. Lampe, "Underwater localization with time-synchronization and propagation speed uncertainties," *IEEE Transactions on Mobile Computing*, vol. 12, no. 7, pp. 1257–1269, July 2013.
6. Z. Gong, C. Li, and F. Jiang, "Passive underwater event and object detection based on time difference of arrival," in *IEEE Global Communications Conference 2020*, Dec. 2020, pp. 495–502.
7. D. Park, K. Kwak, W. K. Chung, and J. Kim, "Development of underwater short-range sensor using electromagnetic wave attenuation," *IEEE Journal of Oceanic Engineering*, vol. 41, no. 2, pp. 318–325, April 2016.
8. J. P. Beaudeau, M. F. Bugallo, and P. M. Djurić, "RSSI-based multi-target tracking by cooperative agents using fusion of cross-target information,"

IEEE Transactions on Signal Processing, vol. 63, no. 19, pp. 5033–5044, Oct. 2015.

9. D. J. Peters, "A Bayesian method for localization by multistatic active sonar," *IEEE Journal of Oceanic Engineering*, vol. 42, no. 1, pp. 135–142, Jan. 2017.

10. Q. Liang, B. Zhang, C. Zhao, and Y. Pi, "TDoA for passive localization: Underwater versus terrestrial environment," *IEEE Transactions on Parallel and Distributed Systems*, vol. 24, no. 10, pp. 2100–2108, Oct. 2013.

11. J. Zheng and Y. C. Wu, "Joint time synchronization and localization of an unknown node in wireless sensor networks," *IEEE Transactions on Signal Processing*, vol. 58, no. 3, pp. 1309–1320, March 2010.

12. Z. Gong, C. Li, and F. Jiang, "AUV-aided joint localization and time synchronization for underwater acoustic sensor networks," *IEEE Signal Processing Letters*, vol. 25, no. 4, pp. 477–481, 2018.

13. M. Aoki, E. Oki, and R. Rojas-Cessa, "Measurement scheme for one-way delay variation with detection and removal of clock skew," *ETRI Journal*, vol. 32, no. 6, pp. 854–862, Dec. 2010.

14. M. Cristea and B. Groza, "Fingerprinting smartphones remotely via icmp timestamps," *IEEE Communications Letters*, vol. 17, no. 6, pp. 1081–1083, June 2013.

15. S. Jana and S. K. Kasera, "On fast and accurate detection of unauthorized wireless access points using clock skews," *IEEE Transactions on Mobile Computing*, vol. 9, no. 3, pp. 449–462, March 2010.

16. M. A. Spirito, "On the accuracy of cellular mobile station location estimation," *IEEE Transactions on Vehicular Technology*, vol. 50, no. 3, pp. 674–685, May 2001.

17. N. Patwari, A. O. Hero, M. Perkins, N. S. Correal, and R. J. O'Dea, "Relative location estimation in wireless sensor networks," *IEEE Transactions on Signal Processing*, vol. 51, no. 8, pp. 2137–2148, Aug 2003.

18. P. Closas and M. F. Bugallo, "Improving accuracy by iterated multiple particle filtering," *IEEE Signal Processing Letters*, vol. 19, no. 8, pp. 531–534, Aug. 2012.

19. P. M. Djurić, J. H. Kotecha, J. Zhang, Y. Huang, T. Ghirmai, M. F. Bugallo, and J. Miguez, "Particle filtering," *IEEE Signal Processing Magazine*, vol. 20, no. 5, pp. 19–38, Sep. 2003.

20. P. M. Djurić, M. Vemula, and M. F. Bugallo, "Target tracking by particle filtering in binary sensor networks," *IEEE Transactions on Signal Processing*, vol. 56, no. 6, pp. 2229–2238, June 2008.

21. S. Han, Z. Gong, W. Meng, C. Li, D. Zhang, and W. Tang, "Automatic precision control positioning for wireless sensor network," *IEEE Sensors Journal*, vol. 16, no. 7, pp. 2140–2150, April 2016.

22. Z. Gong, C. Li, F. Jiang, R. Su, R. Venkatesan, C. Meng, S. Han, Y. Zhang, S. Liu, and K. Hao, "Design, analysis, and field testing of an innovative drone-assisted zero-configuration localization framework for wireless sensor networks," *IEEE Transactions on Vehicular Technology*, vol. 66, no. 11, pp. 10 322–10 335, Nov. 2017.

23. W. Burdic, *Underwater Acoustic System Analysis.* Peninsula Publ., 2002.
24. Z. Gong, C. Li, and F. Jiang, "Analysis of the underwater multi-path reflections on Doppler shift estimation," *IEEE Wireless Communications Letters*, vol. 9, no. 10, pp. 1758–1762, 2020.
25. Z. Gong, C. Li, F. Jiang, and J. Zheng, "AUV-aided localization of underwater acoustic devices based on Doppler shift measurements," *IEEE Transactions on Wireless Communications*, vol. 19, no. 4, pp. 2226 – 2239, Jan. 2020.
26. J. Liang, L. Xu, J. Li, and P. Stoica, "On designing the transmission and reception of multistatic continuous active sonar systems," *IEEE Transactions on Aerospace and Electronic Systems*, vol. 50, no. 1, pp. 285–299, Jan. 2014.
27. J. Zhao, W. Xi, Y. He, Y. Liu, X. Y. Li, L. Mo, and Z. Yang, "Localization of wireless sensor networks in the wild: Pursuit of ranging quality," *IEEE/ACM Transactions on Networking*, vol. 21, no. 1, pp. 311–323, Feb 2013.
28. R. van Vossen, S. P. Beerens, and E. van der Spek, "Anti-submarine warfare with continuously active sonar," *Sea Technology*, vol. 52, no. 11, pp. 33–35, 2011.
29. N. F. Josso, C. Ioana, J. I. Mars, and C. Gervaise, "Source motion detection, estimation, and compensation for underwater acoustics inversion by wideband ambiguity lag-doppler filtering," *The Journal of the Acoustical Society of America*, vol. 128, no. 6, pp. 3416–3425, 2010.
30. L. B. Almeida, "The fractional Fourier transform and time-frequency representations," *IEEE Transactions on Signal Processing*, vol. 42, no. 11, pp. 3084–3091, Nov. 1994.
31. H. M. Ozaktas, O. Arikan, M. A. Kutay, and G. Bozdagt, "Digital computation of the fractional Fourier transform," *IEEE Transactions on Signal Processing*, vol. 44, no. 9, pp. 2141–2150, Sep. 1996.
32. Z. Gong, C. Li, and F. Jiang, "A machine learning-based approach for auto-detection and localization of targets in underwater acoustic array networks," *IEEE Transactions on Vehicular Technology*, vol. 69, no. 12, pp. 15 857–15 866, Dec. 2020.

Chapter 9

Smart Spectrum Usage

Lina Pu[1,*], Yu Luo[2] and Zheng Peng[3]

[1] *University of Alabama, Tuscaloosa, AL 35487-0290, USA.*
* *Lina Pu is the corresponding author.*

[2] *Mississippi State University, Mississippi State, MS 39762, USA.*
[3] *College University of New York, New York, NY, 10031, USA.*

Email: lina.pu@ua.edu, yu.luo@ece.msstate.edu, zpeng@ccny.cuny.edu

9.1. Introduction

As a benefit from its long propagation range, sound is widely used by both "natural acoustic users" (i.e., marine animals) and "artificial acoustic systems" (e.g., sonars, acoustic modems, fish finders, etc.) for communications, ranging, and target detection. Ocean environment features high competition for acoustic spectrum. Meanwhile, available acoustic frequencies in oceans are quite limited, due to the severe frequency-dependent attenuation and the narrowband response of acoustic transducers [1]. Therefore, the spectrum is a scarce resource for underwater acoustic systems.

Figure 9.1 shows a typical underwater environment where multiple acoustic systems coexist. This target detection UAN consists of bottom nodes and surface nodes as well as AUVs collecting sensing results via an ad hoc network whenever the presence of a target is detected, then keeping quiet for the rest of time. Each of bottom nodes equips with specific sensors to monitor the underwater environment, e.g., the existence of marine mammals and ships. AUVs can bring the collected data back by cruise between smart ships and bottom nodes periodically.

Due to the ever-increasing human activities in the ocean, anthropogenic noise has significantly altered the soundscape of the ocean. The negative

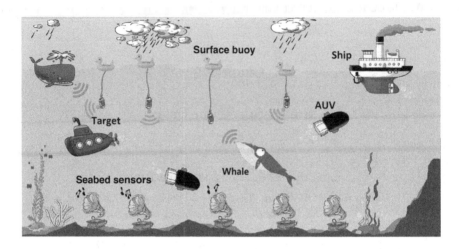

Fig. 9.1.   An example of UAN target detection [2]

impacts of anthropogenic noise on marine animals have been observed for years.   Anthropogenic noise not only induces physiological and behavioral changes in marine animals (marine mammals, fishes, and invertebrates) [3], [4] but also drives them to abandon their habitats and increases their mortality [5]. Different from pollutants that will have a persistent impact on marine ecosystems, the anthropogenic noise can be removed swiftly by eliminating sources.

With a quieter propeller and thrusters, we can reduce the passive marine noise from vessels and limit its impact to a local area. However, the active acoustic signals from acoustic modems are designed to travel far. If we keep greedy usage of the acoustic spectrum, the rapid growth of acoustic devices and traffic explosion in underwater communication networks may cause severe and irreversible damage to the marine ecology.

Following these concerns, underwater CA communication is advocated as a promising solution for smart spectrum usage to achieve the *environment-friendly* and *spectrum-efficient* transmission over acoustic channels [1], [6], [7]. Similar to Cognitive Radio (CR), CA users are allowed to intelligently detect whether any portion of the acoustic spectrum is vacant, and correspondingly change their transmission frequencies, power, or other operating parameters to temporarily use the idle frequency for communications without interfering with other acoustic systems.

9.2. Practical Issues in Underwater Communications

9.2.1. *Heavily Shared Acoustic Spectrum*

In a terrestrial wireless network, the transmission of a radio signal does not affect the communication of humans, birds, and other animals, since the voice signal and the radio signal use different waves to carry the information, which does not interfere with each other. On the other hand, marine animals, UANs and sonars have to share the precious acoustic spectrum with each other, since they all use sound signals for communications. Figure 9.2 shows the overall bandwidth associated with different acoustic systems.

Taking a closer look at Figure 9.2 we observe that from the spectrum point of view, the underwater mid-frequency band is heavily shared. Marine mammals use this frequency band for orientation, communication, and foraging [8], while sonars transmit on this frequency band for navigation and bathymetry.

To name a few, toothed whales communicate on frequencies around 10 kHz; the echolocation signal produced by killer whales is 12 to 25 kHz; the whistle signal (for communication) and click signal (for echolocation) sent by bottlenose dolphin are 200 Hz to 24 kHz and 200 Hz to 150 kHz, respectively. The sea lions could hear the sound frequency up to 70 kHz, and vocalize from 100 Hz to 10 kHz.

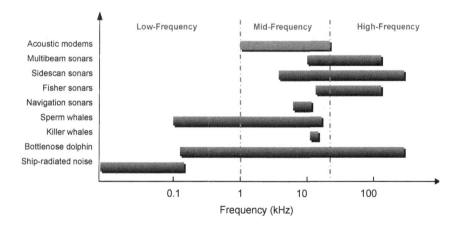

Fig. 9.2.   Spectrum usage of different acoustic systems [1]

Moreover, the operating frequency of sonar systems varies from hundreds of hertz to hundreds of kilohertz depending on the requirement of applications. Specifically, the frequency of CW signal is usually 8 to 16 kHz for navigation and ranges from 10 kHz to hundreds of kilohertz for bathymetry, e.g., 12 kHz in Simrad EM120 and 32 kHz in Simrad EM300. The fishery sonar, which is widely employed to find and harvest fish, works on frequencies from 20 to 200 kHz. These sonars usually have a high source level from 185 to 200 dB re 1 μPa, thereby causing strong interference on surrounding acoustic systems.

9.2.2. *Long Preamble Sequence*

For incoming signal detection, Automatic Gain Control (AGC), and channel estimation purposes, a modem needs to add a sequence of specific signals, called the preamble, before each packet.

In a radio network, the duration of a preamble signal is very short, usually less than several hundreds of microseconds. For example, the preamble in IEEE 802.20 for MBWA consists of 8 symbols, and 104 μs for each one, i.e., total 832 μs preamble sequence [9]. In IEEE 802.22 for cognitive WRAN, the preamble segment includes three components: a superframe preamble, a frame preamble, and a coexistence beacon protocol preamble, and the total length of them is less than 1 ms [10]. By contrast, the preamble signal in acoustic communications could reach one second or even longer, three orders of magnitude larger than that in radio networks. One example is AquaSeNt high-speed Orthogonal Frequency-Division Multiplexing (OFDM) Modem [11]. The packet always consists of two preamble blocks and an OFDM data block, no matter how few data bytes need to be transmitted. In this way, the minimum packet transmission time is the transmission time of two preamble blocks plus a data block, whose waveform is illustrated in Figure 9.3. As shown in Figure 9.3, the preamble takes the first two blocks for packet detection and synchronization, and the third block is the actual data part. From Figure 9.3, we can observe that the minimum packet transmission time in the AquaSeNT OFDM modem is longer than 0.5 seconds.

Table 9.1 lists the length of the preamble sequence in three different acoustic modems. Two folds result in the long preamble of acoustic modems: the low data rate of acoustic modem and the long multi-path of underwater channel are introduced below.

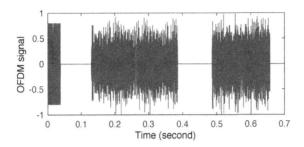

Fig. 9.3.   Preamble significantly elongates the length of a packet

Table 9.1.   Transmission rate and preamble length of different acoustic modems [1]

Modem	Data Rate	Preamble Length
Benthos ATM-88X	140 - 2400 bps Multi Frequency Shift Keying (MFSK) 2.56 - 15.4 kbps Phase-shift keying (PSK)	1.2 s
AquaSeNt OFDM	3 kbps (1/2 coding rate, 4 Quadrature amplitude modulation (4QAM)) 9 kbps (3/4 coding rate, 16QAM)	0.5 s
WHOI Micro	80 bps (Standard) 300 - 5000 bps (High PSK mode)	0.87 s

I) Low data rate of acoustic modems: It is the primary reason
for the long preamble in acoustic modems. Specifically, one major task of
a preamble signal is to work as a mark, indicating the accurate time to
receive oncoming data segments. We call it point-to-point synchronization.
To achieve a good synchronization performance, a sequence with hundreds
of known bits, such as a PN sequence, is usually attached in a preamble. In
radio networks, this PN sequence could be sent out in a very short time. In
an underwater environment, however, the low data rate of acoustic modems
(Table 9.1), extends the transmission time significantly. Taking 512-bit PN
signal as an example, a node in IEEE 802.22 with 22.69 Mbps transmission
rate takes only 22.6 μs to send it out, but the transmission time is extended
to 0.64 secs for an acoustic modem with 800 bps data rate.

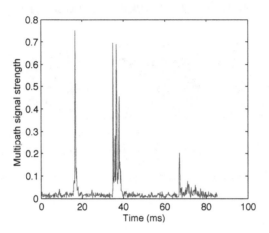

Fig. 9.4.   Long multi-path of acoustic channel

II) Long multi-path of the underwater channel: This is another factor that contributes to a long preamble in UANs. Specifically, a preamble signal is usually comprised of several blocks, and each block serves different functionalities. To overcome the inter-block interference in multi-path environments, a guard time for PSK and FSK-based modulation scheme, or a cyclic prefix for OFDM-based modem needs to be inserted between these blocks. The length of a guard time or a cyclic prefix depends on the length of the multi-path. The radio channel has a very short multi-path owing to the high propagation speed of electromagnetic signals. In this circumstance, the guard time or the cyclic prefix could be as short as tens of microseconds (e.g., 4.7/16.7/53.3 μs in 3GPP LTE standard on different channel conditions). By contrast, the multi-path could reach tens of milliseconds or even longer in underwater communications, as shown in Figure 9.4, depending on the network deployment and channel conditions. The length of the guard time or cyclic prefix signal, in this case, is considerably increased by almost 1000 times longer than that in radio networks.

9.2.3. *Heterogeneous Acoustic Channel*

Due to the heterogeneous geometry of sea surface and seabed, the diverse multi-path effect could result in greatly different communication quality through the network. We had observed this phenomenon when we towed the modems around trying to get reliable communications. Good links on N3, N7, and N8 had less than 15% loss ratio in our observation, compared

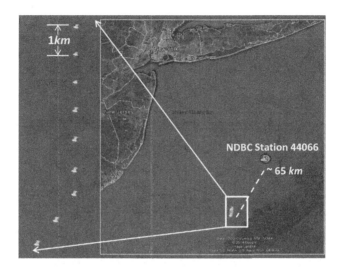

Fig. 9.5.   Deployment of Atlantic Sea test

with almost 65% packet loss with bad channels on N5. Here Figure 9.6 came from a randomly chosen test. Even though packet loss ratios were not the same among other tests, similar diversity on packet loss ratios was observed.

Besides the substantially varied loss ratio on different nodes, the packet delivery features heterogeneous when packets travel in different directions. Taking the network in Figure 9.5 as an example, we call the links for southern directional communication as forward links and the reverse ones as backward links. In Figure 9.6, the forward links suffered more severe packet losses than the backward links when $N4$ and $N5$ were sending. On the contrary, the forward links for $N2$, $N3$, and $N6$ had better reliability than the

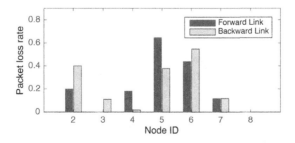

Fig. 9.6.   Heterogeneous packet loss ratio

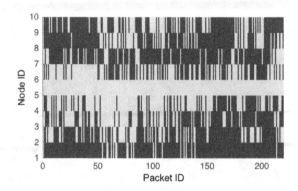

Fig. 9.7.   Dynamic transmission range.

reverse links. This significantly varied delivery means that packets traveling in different directions can have dramatically different packet loss ratios, and therefore brings trouble to MAC protocols relying on homogeneous network assumption.

Under a combined impact of the broadcast nature of the acoustic signal and the unstable underwater channel condition, the communication range not only varies spatially, but also shows dynamic nature in the temporal dimension. Owing to the time-varying nature of wind, current, marine mammal noise, and man-made activities, the link reliability feature changes with time. Figure 9.7 illustrates the dynamic communication range when the middle node (N5) sent packets of 200 bytes. The x-axis on the graph is the index of transmitted packets ordered by sending time, which stands for the time records. The y-axis gives the transmission range in the unit of hops. The positive and negative Hop IDs represent the transmission in the two directions. Packets were sent from Hop 0 in this figure. We highlight the region where nodes successfully received packets from the sender. As shown in Figure 9.7, the transmission range has a remarkable variation with time. In some periods, no packets could be reliably delivered to any node further than one hop away. On the contrary, in the rest of the time, the sender had good communication reliability for transmissions in both directions.

The heterogeneous deliveries across the network are a result of the complicated underwater environments. For acoustic modems like Benthos ATM-885 using FSK-based modulation scheme, the severe multipath effect is one of the most fundamental obstacles to robust underwater communications. How to deal with the diverse packet deliveries among the networks will be a big challenge on the practical spectrum sharing in UANs.

9.3. Smart Spectrum Sharing

Underwater environment, where multiple networks coexist, usually features high competition for acoustic spectrum (channel) among different users. Meanwhile, available communication frequencies in water are quite limited, due to the severe frequency-dependent attenuation and the narrowband response of an acoustic transducer. Therefore, the spectrum is a scarce resource for underwater acoustic systems.

To improve the efficiency of spectrum utilization in a complex underwater environment, underwater cognitive acoustic is advocated as a promising technique to achieve both environment-friendly and spectrum-efficient communications over acoustic channels [1], [6], [7].

In CA, users are capable of sensing the surrounding environment first, and then dynamically configure their operating frequency, transmission power, or other system parameters to avoid interference with other acoustic systems. Figure 9.8 shows the three major components of a CA system: a spectrum sensing mechanism, a dynamic power control algorithm, and a spectrum management system.

- *Spectrum sensing mechanism:* This mechanism plays a crucial role in detecting the presence of other acoustic users and identifying the idle channel for CA. Spectrum sensing could be performed in frequency, time, space, and code domains. Also, users can sense the spectrum

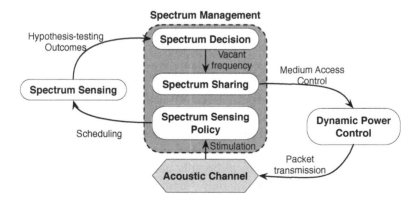

Fig. 9.8.   Architecture of CA system.

independently to increase the channel access opportunity, or coopera-
tively to improve the sensing accuracy.

- *Dynamic power control*: In CA, users could adopt a dynamic power control algorithm to improve the channel capacity and energy efficiency of
  the network. This could be done by assigning proper power to each user
  on different channels while maintaining a constant energy consumption
  over the whole spectrum.

- *Spectrum management system*: It allows acoustic nodes to intelligently
  detect whether any portion of the acoustic spectrum is vacant, and
  change their transmission frequencies, power, or other operating parameters to temporarily use the idle frequency for communications without
  interfering with other acoustic systems.

Next, we will introduce two representative smart spectrum sharing solutions for efficient and environment-friendly spectrum utilization in smart
ships.

9.3.1. *Spectrum Sensing: Unique Challenges and Solutions*

Generally speaking, a CA node may not be able to sense all frequencies in
one sensing period, since a full-band spectrum sensing is not only energy and
time inefficient but also hardware demanding, which makes it impractical
for battery-powered underwater equipment. Thereafter, we consider the
scenario that each CA node could only sense one or several subset frequency
bands in one sensing period. In an asynchronous network, when a node
is sensing the spectrum, other senders may be transmitting on the same
channel, which will interfere with the sensing process. CA users are thus
required to distinguish signals of marine animals from that of CA nodes,
which is a unique challenge of spectrum sensing in cognitive acoustics.

Here, we advocate *cyclostationary* based spectrum sensing approaches
to achieve this goal. Different man-made communication signals naturally
have cyclostationary features at different cyclic frequencies [12]. By recog-
nizing the cyclostationary pattern during spectrum sensing, CA nodes can
distinguish between received signals from different systems.

However, one objective of CA is to share the acoustic spectrum with
marine animals in an environment-friendly manner. Hence, Primary User
(PU) in oceans may involve not only "artificial acoustic systems", such
as UANs and sonars, but also "natural acoustic systems", such as whales
and dolphins. One important question coming up is that whether signals

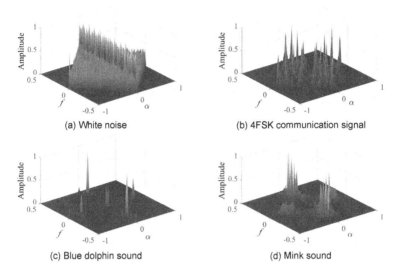

(a) White noise   (b) 4FSK communication signal

(c) Blue dolphin sound   (d) Mink sound

Fig. 9.9. Normalized cyclic cross-periodogram of different acoustic signals in oceans with Hamming window

from "natural acoustic systems" and from CA nodes can be told apart by applying a cyclostationary-based sensing technique. Here, we compare the cyclostationary-based time-smoothed cyclic cross-periodogram [13], $S_x^\alpha[k]^*$, of different acoustic signals in Figure 9.9. The results may help us to answer this question.

Figure 9.9 includes 4FSK signal, which is common in man-made communication systems, and voice signals from two different marine mammals. This figure illustrates that the ambient noise does not exhibit the cyclostationary feature, since its cyclic cross-periodogram has no peaks if $\alpha \neq 0$. By contrast, both 4FSK and voice signals from marine mammals exhibit cyclostationary features on multiple cyclic frequencies. Moreover, it is easy to observe the obvious differences among cyclostationary patterns of different signals (i.e., peaks at different α, where $\alpha \neq 0$). A node could use the position of these peaks to identify sensed signals.

$^*S_x^\alpha[k] = \frac{1}{D}\sum_{d=0}^{D-1} X_l[k]X_l^*[k-\alpha]W[k]$, where α, D, M, $W[k]$ and $X_l[k]$ are the cyclic frequency, the number of windows, the number of samples, the smoothing spectrum window and the Fourier transform of the sensed signal $x[n]$, respectively.

9.3.2. *Receiver-Initiated Spectrum Management*

The smart ship senses the ocean world through underwater sensors, which can communicate with each other to help the smart ship make intelligent decisions. The smart ship collecting data through multi-hop communications naturally forms a centralized data gathering network. Receiver-Initiated Spectrum Management (RISM) is such a receiver-initiated approach that can leverage the data-gathering feature in smart ship communications.

In RISM, the intended receiver first schedules a sensing pattern, i.e., which frequencies senders should work on, for its neighboring senders. Thereafter, by collecting local sensing results from its neighbors, a receiver will have a global picture of the spectrum usage. Finally, for high throughput and low delay purposes, the receiver assigns vacant frequencies and optimal transmission power for its surrounding senders based on the spectrum sensing results and the quality of acoustic links. RISM system allows smart ships to efficiently and friendly share the precious spectrum resource with both "natural acoustic systems" and "artificial acoustic systems".

9.3.2.1. *RISM protocol description*

RISM is a "semi-centralized" system, where receivers run as "semi-center" collecting local sensing results and performing channel allocation for surrounding senders for efficient and reliable data communications. Control packets, which are used for the negotiation among CA nodes to avoid collisions, could be easily shared by the collaborative spectrum sensing and resource (channel and power) allocation without generating extra traffic loads. Therefore, the spectrum sensing mechanism, the spectrum sharing scheme, and the spectrum decision algorithm could be considered as a whole in the new system.

In RISM, the handshake process is initiated at the receiver side to negotiate the vacant spectrum sharing. It involves six phases, as shown in Fig. 9.10.

Phase 1: The receiver broadcasts a Request-To-Receive (RTR) packet to start a handshake process. Here, the RTR message has three functions: (a) to request data, (b) to arrange the transmissions of Available-To-Send (ATS) packets from senders, and (c) to schedule spectrum sensing for senders.

Phase 2: The invited senders will first sense the frequency as arranged and respond with ATS messages to establish connections with the receiver. In order to avoid collisions among ATS packets, senders transmit

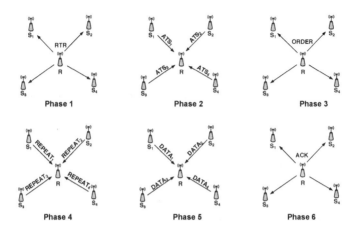

Fig. 9.10. Six phases of the RISM scheme (R is a receiver and s_i is sender i)

ATS following the schedule ordered in RTR. Here, the ATS has four functions: (a) to establish a connection with the receiver, (b) to inform the receiver of its number of data packets to send, (c) to notify the receiver about the spectrum usage of other cognitive users for collision avoidance, and (d) to upload the local sensing result for a collaborative PU detection.

Phase 3: After ATS packet reception, the receiver aggregates local sensing outcomes for the final spectrum decision. Then it broadcasts an ORDER packet, which includes information about the frequency allocation and the transmission power assignment for its neighbors.

Phase 4: If a sender successfully receives the ORDER message, it extracts its schedule information and broadcasts this information to avoid a data collision with other receivers.

Phase 5: After the transmission of REPEAT, each sender sends out its DATA packet at the scheduled time according to ORDER received in Phase 3.

Phase 6: Finally, for reliable transmission, the receiver replies a common acknowledgment (ACK) after the data reception.

Here, RTR, ATS, ORDER, REPEAT and ACK are all control packets and thus, share the Common Control Channel (CCC). From the above description, it is easy to obtain that, though there are six phases in RISM, each round of negotiation allows multiple senders to reserve the channel. Also, the receiver can effectively work as a fusion center to schedule the sensing pattern and to collect local sensing results for collaborative PU detections, and it

could also play a role as the control center to arrange the data transmission
of its surrounding senders.

9.3.2.2. *Throughput optimization in spectrum decision*

When assigning vacant frequencies for communication, receivers need to
schedule the data transmission for its intended senders carefully to avoid
potential collisions. Also, CA nodes in an underwater environment usually
experience severe frequency selective fading. However, a frequency that
appears in deep fading to a node may be of good quality for other nodes.
Therefore, a receiver should dynamically allocate the frequency and power
to senders based on their channel situations.

In a communication system, if the instantaneous Channel State Information (CSI) is available, the receiver could schedule sender k to transmit
at the maximum rate $R_{nk}^t = C_{nk}^t$. Here, C_{nk}^t is the channel capacity of node
k on channel n at time t, which is expressed as

$$C_{nk}^t = a_{nk}^t B_n \log_2(1 + \frac{p_{nk}^t |h_{nk}^t|^2}{N_0 B_n a_{nk}^t}), \tag{9.1}$$

where p_{nk}^t is the transmission power, h_{nk}^t is the instantaneous channel gain
between sender k and the receiver on channel n, B_n is the bandwidth of
channel n, and N_0 is the noise spectral density. The frequency allocation
matrix at time t is represented as $\mathbf{A}^t$, where each entry $a_{nk}^t \in \{0, 1\}$. If
channel n is assigned to node k, then we set $a_{nk}^t = 1$, otherwise $a_{nk}^t = 0$.

However, the real-time channel gain is usually unavailable due to the
long propagation delay and the high dynamic of an underwater channel.
Therefore, we use the *outage probability*[†], which only requires the statistical knowledge of h_{nk}^t, to calculate the channel capacity. The statistical
information of h_{nk}^t is stable and easy to get in underwater communications [1].

Depending on the Quality of Service (QoS) requirement, the packet loss
ratio between sender k and the receiver on channel n should be equal to or
less than a predetermined outage probability β_{nk}. That is

$$Pr[R_{nk}^t > C_{nk}^t] = \beta_{nk}, \tag{9.2}$$

where R_{nk}^t is the data transmission rate.

[†]If one packet is transmitted with spectral efficiency $S(\rho)$ (in bit/sec/Hz) and SNR ρ,
the probability that this packet will be correctly decoded is $1-\beta$, i.e., $P_e(\rho, S(\rho)) = \beta$.
Then $P_e(\rho, S(\rho))$ is called the outage probability [14].

Assume that $|h_{nk}^t|$ follows Rayleigh distribution, then its Probability Distribution Function (PDF), $f(|h_{nk}^t|^2)$, is an exponential distribution with mean value λ_{nk}. Substituting (9.1) into (9.2) and using the exponential expression of $f(|h_{nk}^t|^2)$, we have

$$R_{nk}^t = a_{nk}^t B_n \log_2 \left[1 + \frac{p_{nk}^t \lambda_{nk} \ln(\frac{1}{1-\beta_{nk}})}{N_0 B_n a_{nk}} \right], \tag{9.3}$$

which will be used for the optimal spectrum decision.

In RISM, senders inform the receiver, through ATS messages, of how many data packets to be sent out. Different nodes may have varied sending requests in each transmission. Now, let Q and T_r denote the total bits of data a receiver will receive and the time spent on receiving these data, respectively. Then, we have

$$Q = \int_0^{T_r} \sum_{n=1}^N \sum_{k=1}^K R_{nk}^t \, dt, \tag{9.4}$$

where R_{nk}^t is the assigned data rate to sender k on channel n at time t.

We aim to minimize T_r in (9.4), which is equivalent to maximizing $\sum_{n=1}^N \sum_{k=1}^K R_{nk}^t$ as Q is fixed. Therefore, we formulate the joint power and frequency allocation as the following optimization problem:

Prob.1 $\qquad \arg \max_{\substack{p_{nk}^t > 0 \\ a_{nk}^t \in \{0,1\}}} \sum_{n=1}^N \sum_{k=1}^K R_{nk}^t,$

where $\quad R_{nk}^t = a_{nk}^t B_n \log_2 \left[1 + \frac{p_{nk}^t \lambda_{nk} \ln(\frac{1}{1-\beta_{nk}})}{N_0 B_n a_{nk}^t} \right].$

s.t. $\tag{9.5}$

C1: $\sum_{k=1}^K a_{nk}^t = 1, \quad n \in \{1,\ldots,N\},$

C2: $\sum_{n=1}^N p_{nk}^t \leq P_k, \quad k \in \{1,\ldots,K\},$

C3: $a_{nk}^t = 0,$ if $c_{nk}^t = 1, \ n \in \{1,\ldots,N\}, k \in \{1,\ldots,K\}.$

In Prob.1, C1 is the channel allocation constraint which ensures that each channel is assigned to no more than one CA sender; C2 is the power constraint to guarantee that the overall transmission power of each sender does not exceed the maximum power supply, and C3 is the collision avoidance constraint.

Prob.1 can be solved by a similar approach proposed in [15] and [16]. We first relax the requirement $a_{nk}^t \in \{0,1\}$ to allow a_{nk}^t to be a real number within the interval $[0, 1]$. Then, the objective function of the problem follows the form of $f(x, y) = \mu x \log_2(1 + \nu y/x)$, where μ and ν are constants. It is easy to prove that the Hessian matrix of this function is negative semidefinite for all x and y. Therefore, the objective function is concave. Finally, Prob.1 is converted to the following classic method of Lagrangian multipliers:

$$
\begin{aligned}
\mathcal{L}(a_{nk}^t, p_{nk}^t) = \sum_{n=1}^{N} \sum_{k=1}^{K} B_n a_{nk}^t \log_2 \left[1 + \frac{p_{nk}^t \lambda_{nk} \ln(\frac{1}{1-\beta_{nk}})}{N_0 B_n a_{nk}^t} \right] \\
- \phi_k \left(\sum_{n=1}^{N} p_{nk} - P_k \right) - \varphi_n \left(\sum_{k=1}^{K} a_{nk}^t - 1 \right),
\end{aligned}
\tag{9.6}
$$

where ϕ_k and φ_n are the Lagrangian multipliers of the constraints C1 and C2, respectively.

By solving the Lagrangian optimization problem, we have

$$
\hat{p}_{nk}^t =
\begin{cases}
0, & \phi_k \geq \dfrac{\theta_{nk}}{\ln(2)} \text{ or } a_{nk}^t \neq 1 \text{ or } c_{nk}^t = 1, \\[2ex]
\min \left\{ \dfrac{B_n}{\phi_k \ln(2)} - \dfrac{B_n}{\theta_{nk}}, \, P_k \right\}, & \text{otherwise.}
\end{cases}
\tag{9.7}
$$

and

$$
k' = \arg\max_k \frac{p_{nk}^t \theta_{nk}}{B_n}, \quad k \in \{1, \ldots, K\}.
\tag{9.8}
$$

Here

$$
\theta_{nk} = \frac{\lambda_{nk} \ln(\frac{1}{1-\beta_{nk}})}{N_0}.
\tag{9.9}
$$

Let vector $\mathbf{A}_k^t$ be the channel assignment to sender k at time t, which is a collection of row indexes with nonzero elements in the kth column of $\hat{\mathbf{A}}^t$. Substituting (9.7) into constraint C2 of Prob.1, we have

$$
\phi_k = \frac{\sum_{n \in \mathbf{A}_k^t} B_n}{\ln(2) \left[P_k + \sum_{n \in \mathbf{A}_k^t} \dfrac{B_n}{\theta_{nk}} \right]}.
\tag{9.10}
$$

Finally, we get the optimal transmission power, $\hat{p}_{nk}^t$, by substituting (9.10) into (9.7).

With the constraint condition C3 considered, we can use the following iterative algorithm to compute $\hat{p}_{nk}^t$ and $\hat{a}_{nk}^t$ in RISM.

Algorithm 1

Initialization: Based on the information of spectrum usage collected from ATS and REPEAT packets, generate the collision avoidance matrix $\mathbf{C}^t$.

Iterative Calculations:
do
   for $n = 1$ to N

      Step 1: Let $a_{nk}^t = 1$ for each sender, k, in turns if $c_{nk}^t \neq 1$, and calculate ϕ_k according to (9.10). In this step, previous channel allocation $a_{n'k}$, $n' \neq n$ remains unchanged.
      Step 2: Use (9.7) to calculate p_{nk}^t.
      Step 3: Pick out the best sender, k', based on (9.8), and set $a_{nk'}^t = 1$.

   end for

   Step 4: Calculate $\sum_{n=1}^{N}\sum_{k=1}^{K} R_{nk}^t$ in Prob.1.

while The increment of $\sum_{n=1}^{N}\sum_{k=1}^{K} R_{nk}^t$ is larger than a predetermined threshold.

After running Algorithm 1, the receiver attaches the sending time of data packets, the channel assignment, transmission power, and data rate into an ORDER, and delivers the packet to intended senders.

9.3.2.3. *Adaptive polling in RISM*

In receiver-initiated approaches, there is a unique challenge that receivers need to decide when to poll the neighboring senders blindly. It becomes a big problem in a distributed network, since receivers usually lack in the current status, e.g., having packets to send or not, of its intended senders.

Adjusting the polling frequency of a receiver will cause a trade-off between the queuing delay and the energy efficiency. Initiating handshake over frequently results in resources (spectrum, energy, and time) waste for transmitting control messages, whereas slowing down the polling rate occasionally leads to larger queuing delay such that the data cannot be delivered timely. In a receiver-initiated protocol, a receiver should adjust its polling frequency, i.e., the time interval between successive RTR requests, to match the traffic loads on its intended senders.

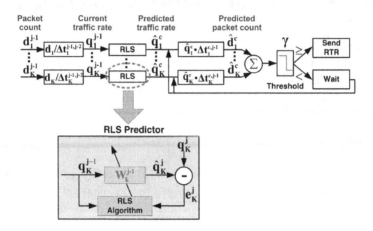

Fig. 9.11. RLS-based adaptive polling scheme for RISM system

RISM adopts the traffic prediction to implement the smart polling scheme. By using the traffic predictor, a receiver could estimate the current traffic loads of its intended senders through historical traffic measurements. Whenever the total traffic of neighboring senders exceeds a threshold, the receiver sends out an RTR packet to request data.

In the literature, a number of algorithms, such as the adaptive filter [17] and artificial neural network [18], have been proposed for predicting the network traffic. In RISM, we could choose an appropriate method depending on the demand of an application. For instance, in a target detection network, the traffic load of a sensor node may change quickly with the entering and leaving of a target. In this scenario, a receiver could employ a Recursive Least Squares (RLS) filter, which has a fast convergence, to track the traffic variation of its neighbors. Moreover, if the network traffic is non-linear, non-stationary, and non-Gaussian but changes slowly, a receiver could use a Finite-Impulse-Response Artificial Neural Network (FIR-ANN) [19] for traffic prediction.

Figure 9.11 shows an example of an RLS filter-based smart polling scheme for the RISM system. By adjusting the threshold γ in the traffic predictor, we could achieve a tradeoff between the queuing delay and the energy efficiency. More specifically, with a small γ, the data produced by each sender could be sent out in a timely manner; whereas, with a large γ, each round of communication could carry more data packets, which improves the energy efficiency.

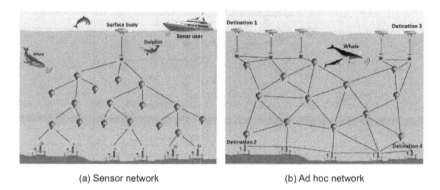

(a) Sensor network   (b) Ad hoc network

Fig. 9.12.   Two network topologies used in simulations. (a) In the sensor network, 20 underwater sensors in the network deliver their data to the sink node (i.e., the surface buoy). (b) In the ad hoc network, 16 underwater nodes randomly select a node at four corners as the destination for data transmission

9.3.2.4. *Performance of RISM*

The performance of RISM has been evaluated in Aqua-Sim [20], an NS-2 based underwater network simulator. Two typical network topologies, as shown in Figure 9.12, have been applied. The sensor network is usually applied for sensing data collection. In the sensor network, we have 20 sub-sea nodes collecting and forwarding data to the sink node on the surface of the oceans, as illustrated in Figure 9.12(a). In the ad hoc network, four hydrophones are deployed at four corners as sink nodes, as shown in Figure 9.12(b). In the network, 16 sub-sea nodes generate and deliver data to a randomly selected hydrophone. In both topologies, the size of data packets is 250 B.

The average distance among neighboring nodes is 1 km (uniformly distributed between 800 m to 1200 m) in both topologies with maximum transmission range and the maximum transmission power of each node as 1.5 km and 20 W, respectively. The overall available channel bandwidth in the simulation is 30 kHz (from 1 to 31 kHz), which is evenly divided into six sub-bands. The lowest sub-band (1 to 6 kHz) is the CCC. The remaining five sub-bands are used as a data channel. In the network, two PUs randomly select one amongst five data channels for its communication and switch the communication channel every 60 secs.

Although RISM aims to eliminate the collisions caused by the hidden terminal problems, we can observe the existence of collisions in RISM when (a) the node fails to overhear the transmission scheduling of surrounding

Smart Ships

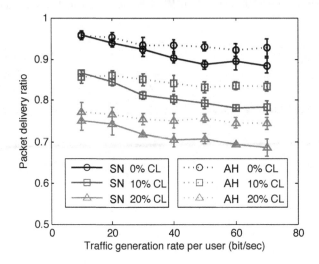

Fig. 9.13. Packet delivery ratio of RISM (SN stands for sensor network, AH stands for ad hoc network, and CL represents channel loss rate)

users due to the collision of ORDER or REPEAT with other control packets, and (b) the node does not overhear the ORDER or REPEAT packets on time due to the long propagation delay in underwater communications. The delivery ratio of data packets with respect to the variation of a traffic load is presented in Figure 9.13. Here, the packet delivery ratio is defined as the number of packets successfully received by receivers divided by the total packets sent by senders.

Figure 9.13 demonstrates that RISM achieves 90% to 95% delivery ratio when there is no channel loss caused by factors other than collisions. RISM can get a higher delivery ratio, which means lower data collisions, in the ad hoc topology than in the sensor network. In the sensor network topology, the data flow is gradually aggregated to the upper nodes in the network causing higher collision probabilities than the case in the ad hoc topology where the destination of data packets is one of a random node located in the four corners resulting in the scattered data flow. When the probability of packet decoding failure caused by the poor channel quality is 10% or 20% in my simulations, the high channel loss becomes the dominant reason for a low packet delivery ratio, as shown in Figure 9.13. However, it is worth noting that the end-to-end reliability of RISM can be guaranteed by the acknowledgment and retransmission mechanism regardless of the packet loss on the hop-by-hop delivery.

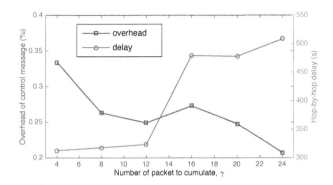

Fig. 9.14. Tradeoff between delay and overhead

In RISM, a tradeoff exists between the packet delivery delay and the overhead of control messages through changing the polling frequency of receivers. More specifically, if a receiver polls its intended senders frequently, data packets on a sender will have a short queuing delay, but at the cost of a low handshake efficiency, since each round of communication carries only a small number of data packets. On the other hand, if a receiver waits for a long time before requesting data, the queuing delay on the senders would be significant. However, a low polling frequency guarantees that enough data packets could be delivered in each round of the handshake process, which improves the efficiency of a negotiation.

For RISM with the smart polling scheme, a receiver would not send RTR until it predicts that the total number of packets accumulated on its intended senders goes over the polling threshold. Therefore, we could change the polling frequency of a receiver by simply adjusting the parameter γ. In Figure 9.14, the traffic generation rate is 40 bit/sec and shows the tradeoff between the overhead of control messages and the hop-by-hop delay of RISM in the sensor network topology with respect to γ. In this figure, the overhead of control messages represents the percentage of energy consumption on control packets for each successful data transmission. It is a ratio of energy consumption on transmitting control messages to that on all packets (control plus data). The results depict that when γ is small, frequent handshake consumes considerable energy on control message transmissions. The energy consumption remarkably reduces with the increment of γ, but results in a larger delivery delay, especially when γ is over 13. Hence, in the rest of this section, the threshold for the adaptive polling is set as $\gamma = 13$ unless stated otherwise.

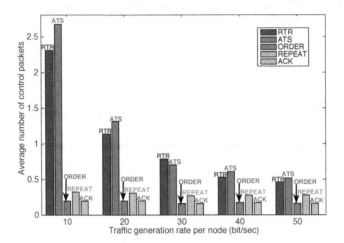

Fig. 9.15. The average number of control packets used for each successful data transmission in the sensor network

Figure 9.15 can give us an insight into the overhead of RISM. From this figure, we could observe a considerable decrease in the average number of control messages with the growth of traffic generation rate. Intuitively, when the traffic is light, the time it takes for senders to cumulate enough data packets for being polled will be inevitably long, which is not desirable in a delay-sensitive application. To tackle this problem, a maximal polling interval can be used. When either the polling interval exceeds the maximal value or the number of cumulated packets reaches γ, the receiver will initiate a request for the data reception. For this reason, in a situation of low traffic rate, a handshake is most likely triggered by the maximal polling interval, at which time senders may have only a few data to send. On the other hand, when the traffic rate is high, the threshold, γ, controls the polling frequency. The average number of control packets used for each data transmission becomes stable as the traffic generation rate grows.

The performance of RISM has been compared with MMAC-CR [21], a representative MAC protocol for CR networks, in two different traffic patterns shown in Figure 9.16. The Poisson traffic is generally used to model the arrival process of traffic in sensor networks where the data traffic is barely bursty. A varying mean value in the Poisson process can simulate the temporal variation of the data collection rate. The Pareto traffic generator could well capture the traffic features of an event-driven sensor network,

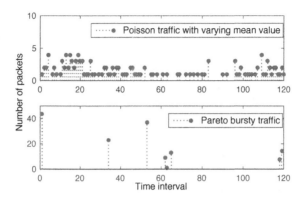

Fig. 9.16. An example of traffic patterns (Average traffic generation rate is 0.02 packet per second for both scenarios. Each time interval is 50 secs.)

where sensor nodes generate a large number of observations whenever a target event is detected.

The performance metric used in the comparison is network throughput, delay, and overhead of control packet. The throughput is bits per second successfully received by each node in a network. The delay consists of queueing delay, transmission delay, and propagation delay. Considering the low throughput and the high collision probability among control messages in UANs, the queuing delay waiting for channel access is considerable and dominates the packet delivery delay. The overhead is calculated as the ratio of energy consumption on transmitting control messages to that on all packets (control plus data).

Figure 9.17 demonstrates the performance comparison in the sensor network (Figure 9.12(a)) where the data generation rate follows Poisson

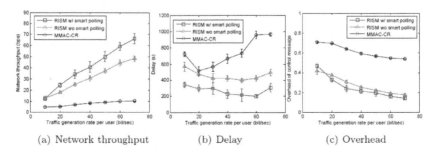

(a) Network throughput   (b) Delay   (c) Overhead

Fig. 9.17. Performance comparison for Poisson traffic with slowly varying mean value in sensor topology

process. From Figure 9.17 we observe that RISM outperforms the conventional sender-initiated MMAC-CR in all aspects. More specifically, smart polling assisted RISM achieves the highest throughput benefiting from the parallel reservation. By allowing receivers to negotiate with multiple senders in parallel, RISM mitigates the problem of a low handshake efficiency caused by the long propagation delay and the long preamble in acoustic modems. Moreover, as shown in Figure 9.17(a), the throughput improvement of RISM dramatically increases with the growth of network traffic generation rate. In applications with heavy traffic loads, RISM could provide high network throughput, which makes it a promising solution for efficient spectrum management. Compared to RISM without smart polling, the traffic prediction scheme grants receivers the capability to dynamically poll senders with the varying mean value of Poisson traffic, thereby leading to a significant throughput enhancement.

Furthermore, RISM with smart polling allows a network to accommodate a high traffic rate with relatively low delivery delays, as shown in Figure 9.17(b). In a network with a heavy traffic load, the packet queuing delay is the main source of packet delivery delays in RISM without smart polling. The traffic prediction enables each receiver to retrieve data from surrounding senders not only more efficiently, but also more timely than the RISM without smart polling, thereby significantly reducing the delivery delay in scenarios with high traffic generation rates.

Another advantage of RISM over sender-initiated protocols is the low overhead on control messages. When receivers start negotiation with their surrounding senders, RISM works as a "semi-centralized" system, where the spectrum sensing, spectrum sharing, and dynamic power control could efficiently share control packets with each other. As illustrated in Figure 9.17(c), RISM with and without traffic prediction has comparable control overhead, since both schemes tend to wait for enough cumulated data packets before starting a handshake process for better energy efficiency. MMAC-CR, by contrast, has almost twice larger overhead than RISM, as MMAC-CR has to schedule separate control messages for spectrum sensing, multi-channel rendezvous, and dynamic power control.

Figure 9.18 uses the same setting as Figure 9.17, but the network topology is changed from the sensor network to the ad hoc network as shown in Figure 9.12(b). By comparing the results in Figure 9.17 and Figure 9.18, we could observe that RISM has lower throughput in the ad hoc network than in the sensor network. This is because, in the sensor network, underwater nodes generate and forward data packets to a common destination,

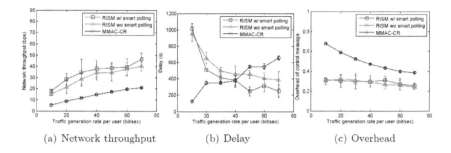

(a) Network throughput   (b) Delay   (c) Overhead

Fig. 9.18. Performance comparison for Poisson traffic with slowly varying mean value in ad hoc topology

namely, the surface node resulting in the aggregated data flow. Therefore, a receiver could easily get a large number of packets from nodes beneath it in each period, which improves the handshake efficiency. In the ad hoc network, on the other hand, the destination of data packets is one of a random node located in the four corners, which "diluents" the traffic. Therefore, a receiver usually retrieves fewer data packets within a given period in an ad hoc network than that in a sensor network, which reduces the handshake efficiency and leads to a lower nodal throughput for RISM.

By contrast, MMAC-CR can achieve higher throughput in ad hoc networks than that in sensor networks. Intuitively, in sensor networks, multiple senders choosing the same relay node causes heavy congestion for the channel access. In an ad hoc network, however, the data packets are scattered resulting in lower collision probability on control messages. Due to a similar reason, RISM has a much longer delay in the ad hoc network than that in the sensor network, whereas MMAC-CR can achieve a shorter delay in the ad hoc network. Although RISM works more efficiently in a sensor network, it still outperforms the sender-initiated MMAC-CR in terms of network throughput and energy efficiency in the ad hoc network.

By comparing the results in Figure 9.17 and Figure 9.19, we could realize how a traffic pattern affects the performance of different protocols. For Pareto bursty traffic, the instantaneous traffic is bursty but the average data generation rate is a constant in the simulation. As for Poisson traffic, the average traffic generation rate varies from 0.5× to 1.5× of the mean value in each test. We observe that RISM with smart polling has comparable performance with both Poisson traffic and bursty traffic, i.e., RISM is barely affected by the traffic pattern. With the assistance of traffic prediction, the

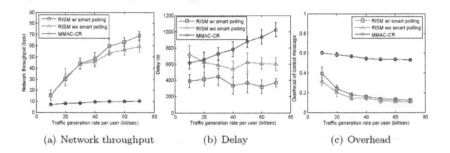

(a) Network throughput   (b) Delay   (c) Overhead

Fig. 9.19. Performance comparison for Pareto bursty traffic in the sensor network

smart polling mechanism could capture the variations in the network traffic, thereby making receivers in RISM request data at a proper time. RISM without smart polling, on the other hand, has a much lower throughput than RISM with smart polling when the network traffic varies with time, as shown in Figure 9.17(a). The difference in throughput of RISM with and without smart polling becomes less significant with bursty traffic. This indicates that adaptive polling could bring more enhancement to RISM when the average traffic rate is more dynamic.

9.3.3. *Spectrum Management with Dynamic Control Channel*

In the previous RISM solution, all the control packets (e.g., RTR, ATS, ORDER, REPEAT, and ACK) are sent over the CCC. The CCC-based MAC protocols are very popular due to their high reliability, easy implementation, and low overhead. However, due to the frequency-dependent attenuation, a CA may not have enough frequency band to its CCC. How to prevent the control channel from congesting in the applications with a heavy traffic load should be considered carefully.

A MAC protocol in a cognitive network relies more on the control packets than that in a conventional single-channel network. Cognitive nodes use these control packets for handshaking to avoid collisions, negotiating to decide the communication channel (multi-channel rendezvous), and collaborating to improve the sensing accuracy (cooperative sensing). In a cognitive network, people usually assign a dedicated CCC for control packets [21], [22], [23]. This channel is physically separated from the in-band channel where data transmission occurs.

However, the available spectrum resource is limited in acoustic communication owing to the frequency-depended attenuation and narrowband response of the acoustic modem. Hence, a CA may allocate a narrow frequency band for control messages, resulting in a congestion problem on CCC in a heavy traffic load situation. This reduces the performance of a cognitive network in terms of throughput, energy efficiency, and end-to-end delay. To solve this problem, some excellent works have been done in recent years.

In this section, we introduce DCC-MAC [2], a Dynamic Control Channel solution for distributed CA networks. The control channel in DCC-MAC consists of two parts: a dedicated CCC and one or more data channels. Once the acoustic nodes detected congestion of CCC, they could flexibly select a proper data channel to extend their control channel and return the excessive frequency bands back when the current control channel becomes idle.

9.3.3.1. *DCC protocol description*

At the beginning of DCC-MAC, a node with a data packet will sense the channel usage of the surrounding environment first. Thereafter, the sender initiates a channel negotiation process through transmitting an RTS message on the control channel to its intended receiver. The RTS packet involves the ID of each vacant channel explored by the spectrum sensing.

If an RTS packet collides with any other control messages, the receiver broadcasts a CPCN to its neighbors after a random backoff; otherwise, it detects whether the oncoming data will collide or not. If not, the receiver senses the channel and selects the one available for both the sender and the receiver. The ID of the selected channel is sent out via a CTS packet.

Once a sender got the CTS, it broadcasts the ID of its following communication channel through an RCTS message on the old control channel. After that, both the sender and its intended receiver turn to the newly selected channel for data transmission. Finally, if the data is successfully decoded, the receiver sends out an ACK packet to the sender for reliable transmission. The flowchart and state machine of DCC-MAC are shown in Figure 9.20 and Figure 9.21, respectively.

Figure 9.22 demonstrates the channel structure of DCC-MAC, which consists of a single CCC dedicated to controlling packets and several in-band channels that are used to send either the control message or the data packet whenever the PU does not occupy it. The in-band channel is further

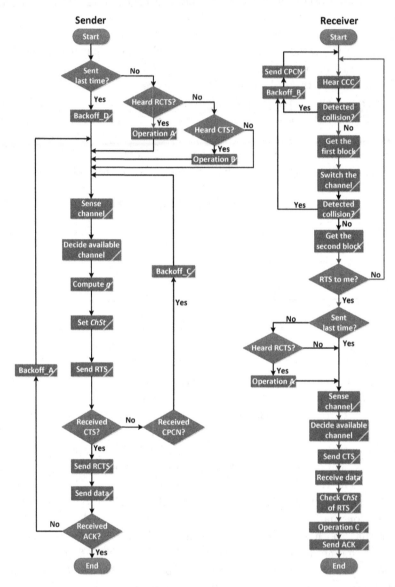

Fig. 9.20. The flowchart of DCC-MAC. Operation A: Remove the selected channels involved in SelCh of RCTS from the available channels; Operation B: Remove the selected channels involved in SelCh of CTS from the available channels; Operation C: Select the channel to increase or decrease the bandwidth of the control channel according to the value if ChSt in the corresponding RTS packet

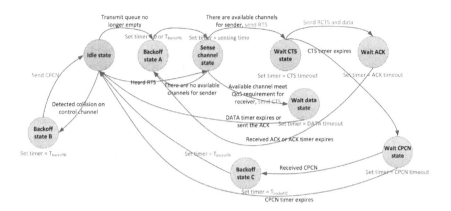

Fig. 9.21.   State machine of DCC-MAC

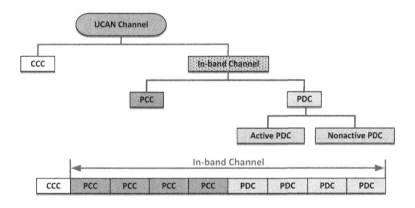

Fig. 9.22.   Channel structure of DCC-MAC

divided into the Preassigned Control Channel (PCC) and the Preassigned Data Channel (PDC). In PCC, a control packet has a higher priority than a data packet to use the channel; in PDC, conversely, the priority of the control packet and that of the data packet are reversed.

In DCC-MAC, a node may choose an in-band channel for data transmission or as a control channel depending on a proper in-band channel selection scheme.

*1) **Send data***: For data transmission, the receiver selects a channel from the vacant PDC with the highest average SNR first. If the selected channel cannot meet the requirement of QoS in terms of the transmission rate and outage probability, more channels are chosen from the idle PDC in descending order of SNR. If there is no vacant PDC, or the QoS still cannot reach the requirement after all available PDCs being selected, the receiver chooses additional channels with high SNR from the nonactive PCC. In order to avoid the collisions between data and control packets, a receiver should not choose the active PCC as the data channel, unless there is neither vacant PDC nor nonactive PCC available.

II) Extend control channel: If a sender and its intended receiver plan to increase the bandwidth of their control channel for their next round of communication, the receiver selects a vacant PCC with the highest average SNR as an extension of the current control channel. If no PCC is available, the node chooses the one from nonactive PCD with the highest average SNR. The receiver should not select an active PDC as the control channel to avoid the collisions between control packets and data packets unless there is neither idle PCC nor nonactive PDC anymore.

III) Remove control channel: If a sender and its intended receiver are going to release the bandwidth of their control channel, the receiver selects an active PDC with the lowest SNR from its current control channel, and then removes it. If there is no active PDC, the receiver removes the nonactive one with the lowest SNR. If the current control channel has neither an active PDC nor a nonactive one, the receiver removes a PCC with the lowest SNR.

Here are some essential rules for channel selection. First of all, due to the high dynamic of acoustic channels [24], the instantaneous receiving SNR may not be available at the sender side; as a result, the average receiving SNR is advocated, which is usually stable in a long period. A node can calculate the average receiving SNRs on a channel based on the historical values it measured from receiving or overhearing control packets and data packets[‡] from its neighbors.

Secondly, when selecting the channel to send data or to extend the bandwidth of the control channel, a node sorts the candidate channels in descending order of the average receiving SNR. Compared to a random selection strategy, this rule helps nodes to use the channel resource more effectively, while reducing the collision probability among data packets. More specifically, due to the spatial diversity [27], the quality of a channel that is good on one node may be bad on another. Therefore, if a node selects

---

[‡]As a critical parameter in communication systems, the receiving SNR can be measured by most of the commercial acoustic modems directly without any modification on the hardware or the software [25, 26].

the channel according to its average receiving SNR on each channel, it has a small chance to compete with multiple neighbors for the same channel. Moreover, choosing a channel with the highest SNR first could get a fast improvement in the transmission rate.

Thirdly, when a receiver decreases the bandwidth of its control channel, it selects the one with the worst average receiving SNR first. There are three advantages of this strategy: (a) the receiver could save the bandwidth for other nodes without losing too much transmission rate of its control messages; (b) due to the spatial diversity of the channel, this strategy gives other nodes an opportunity to efficiently use the removed channel if they have higher receiving SNR on this channel than its current owner, and (c) the receiver can avoid adjusting the bandwidth of its control channel frequently.

9.3.3.2. *DCC performance evaluation*

The performance of DCC-MAC has been evaluated with two network topologies similar to Fig. 9.12. RISM and MMAC-CR are used as benchmark protocols for comparison purposes.

In the simulation, the overall communication bandwidth is 16 kHz (from 17 to 33 kHz), which is evenly divided into 32 sub-channels. The lowest 500 Hz (17 to 17.5 kHz) is assumed not to be occupied by any PUs in the area and is assigned as the CCC. The remaining 31 sub-channels are used as the in-band channel. In the CA network, two PUs randomly select one amongst 31 in-band channels for its communication and switch its communication channel every 100 secs. In the test, each sender generates data packets of size 250 B following the Poisson process. The length of the preamble sequence in the acoustic modem is 0.2 secs.

The effectiveness of the dynamic control bandwidth adjustment in DCC-MAC is demonstrated in Figure 9.24, which presents a snapshot of the varying control bandwidth on nodes N_4 and N_9 in the ad hoc network. The positions of N_4 and N_9 are marked in Figure 9.23. Although both nodes have the same data generation rate on the application layer, they get different traffic loads on the MAC layer due to the data forwarding in the multi-hop network. Besides the self-generated data, N_9 also helps to relay data for neighbors to the destination, N_{10}. On the contrary, N_4 has few data forwarding requests resulting in light traffic load. As a result of different traffic loads on two nodes, the bandwidth of their control channel is apparently different. When the network was stable, N_4 used about 8 kHz

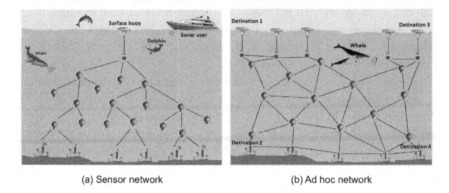

(a) Sensor network   (b) Ad hoc network

Fig. 9.23. Network deployment in simulations. In ad hoc topology, 16 underwater sensors select a random destination at four corners (N_{10}, N_{11}, N_{13} and N_{18}) as the final destination. In the sensor network, 20 nodes deliver their data to node N_0.

as the control channel, and N_9 only occupied roughly 3 kHz for control communication. The spatially and temporally varying bandwidth of the control channel validates the effectiveness of the control adjustment scheme in DCC-MAC.

As a result of the efficient bandwidth adjustment mechanism on the control channel, DCC-MAC outperforms RISM and MMAC-CR in terms of high handshake success rate, as shown in Figure 9.25. The handshake success rate is calculated by the number of transmitted CTS/ORDER/ATIM_ACK dividing the number of RTS/RTR/ATIM sent out, in DCC-MAC, RISM, and MMAC-CR. From this figure, we observe

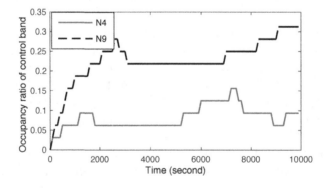

Fig. 9.24. Snapshot of dynamic bandwidth adjustment on control channel

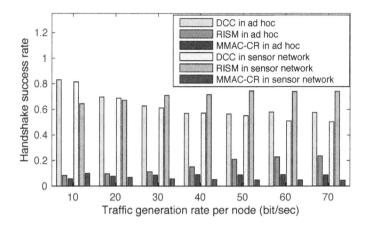

Fig. 9.25. Comparison of success rate of negotiation.

that (a) the competition success rate of DCC-MAC is much higher than MMAC-CR and RISM in an ad hoc topology, and (b) the successful competition rate of RISM increases with a growth of traffic generation rate, but it is sensitive to the network topology.

By smartly adjusting the bandwidth when the control channel is congested, the collisions among the control packet can be reduced significantly. This also explains the comparable performance of DCC-MAC in ad hoc and sensor networks, although they have extremely different traffic patterns. The traffic in ad hoc topology is scattered across the network, whereas the data packets tend to gather in a sensor network. No matter the traffic pattern varies in two networks, the control bandwidth in DCC-MAC can adaptively adjust according to the traffic loads. In contrast, RISM has a very low handshake success rate in ad hoc networks but a high success rate in the sensor network. In ad hoc networks, the inefficient data polling mechanism is the main cause of failed handshakes in the receiver-initiated MAC protocols. In the sensor network, however, the gathering traffic flow can considerably increase the polling success rate and, meanwhile, reduce the control overhead in RISM, resulting in a high handshake success rate. MMAC-CR has the lowest competition success rate in both topologies due to the heavy congestion on the CCC of 3.4 kHz.

Figure 9.26 compares the delivery ratio of the data packet, which is defined as the ratio of the number of data packets successfully received by receivers. We can observe that DCC-MAC has higher data collisions than RISM and MMAC-CR, especially in a sensor network. This is a penalty of

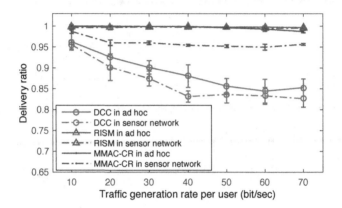

Fig. 9.26.   Comparison of data delivery ratio

the dynamic control channel scheme in DCC-MAC. The data collision can not be avoided completely, even though the random backoff mechanism is designed. Note that in the sensor network the data flow is gradually aggregated to the upper nodes in the network. Hence, the traffic load of a node near the destination (node N_0) may be much higher than the traffic generation rate. In this situation, data packets will have a higher probability to collide with each other if a MAC protocol is not designed for collision-free data transmission. RISM has the highest delivery ratio as receiver-initiated MAC naturally has better data collision capability than conventional sender-initiated DCC-MAC and MMAC-CR. MMAC-CR has a higher delivery ratio than DCC-MAC owing to its two-phase design. Only nodes assigned with the same channel compete for channel access in the data phase of MMAC-CR, resulting in lighter collision probability. In the sensor network, however, the data are gathered and the network becomes "crowd", which leads to a lower delivery ratio for MMAC-CR.

Figure 9.27 demonstrates the network performance of the three protocols in terms of throughput and energy efficiency. From this figure, we observe that DCC-MAC outperforms the other two protocols in almost all aspects. In addition, compared with RISM and MMAC-CR, the performance of DCC-MAC is resilient to the topology of the network.

More specifically, as shown in Figure 9.27(a), DCC-MAC achieves the highest throughput benefiting from its smart bandwidth adjustment mechanism on the control channel. The dynamic control channel allows DCC-MAC to achieve more efficient spectrum utilization. In addition, by

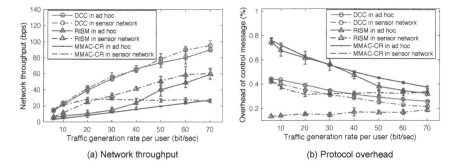

Fig. 9.27.   Performance comparison for Poisson traffic in ad hoc and sensor networks

reducing the collision probability of control messages, the congestion on CCC is not the bottleneck of DCC-MAC to achieve high throughput anymore. By contrast, MMAC-CR has the lowest throughput mainly due to the high collisions on the CCC, which is also revealed in Figure 9.25. Although RISM has a comparable or even higher handshake success rate than DCC-MAC in the sensor network, the lower spectrum efficiency leads to lower throughput. In addition, among the three protocols, DCC-MAC is the only protocol that has comparable network throughput in ad hoc and sensor networks. As most MAC protocols are sensitive to the network topology and traffic load, the feature of DCC-MAC on resilient to different network situation make DCC-MAC a practical solution in various underwater applications.

Figure 9.27(b) compares the overhead, which is defined as the ratio of the energy consumed on transmitting control messages over the total energy consumption (on control plus data packets), of different protocols. This figure reveals that DCC-MAC has the lowest overhead in an ad hoc network benefiting from the lowest collision probability on the control channel. RISM and MMAC-CR have the highest overhead in ad hoc topology for different reasons. Because of the scattering traffic in an ad hoc network, the receivers fail to poll neighboring senders frequently when there are few data cumulated. In MMAC-CR, the heavy collisions on CCC are the main cause of the high overhead. This result is consistent with the handshake success rate illustrated in Figure 9.25. The energy efficiency of DCC-MAC in the sensor network is comparable within ad hoc networks, which is a result of adaptive control bandwidth adjustment to the network traffic load. However, RISM and MMAC-CR are sensitive to the topology of a network. The aggregated traffic in a sensor network significantly improves the polling success rate in RISM and leads to the lowest overhead in the data-gathering network. Although MMAC-CR has higher collisions on the control channel in the sensor network than in ad hoc topology, the packet train in data transmission improves the average energy efficiency.

9.4. Summary

In this chapter, we introduced cognitive acoustic as a solution to the smart spectrum sharing for the smart ship. Two representative cognitive MAC protocols, namely, RISM and DCC-MAC, have been discussed for environment-friendly and spectrum-efficient spectrum sharing.

In RISM, the receiver initiates each round of the handshake process. This strategy allows the receiver in each round of handshake to request packets from multiple senders in parallel. In addition, in RISM, the three components, i.e., the cooperative spectrum sensing, spectrum sharing, and spectrum decision, do not generate control messages separately as independent pieces. Instead, they share control packets with each other without incurring additional control overhead, which significantly improves the negotiation efficiency considering the high latency in CA and the long preamble in acoustic modems.

However, there is a unique challenge in receiver-initiated approaches for receivers to decide when to poll without prior knowledge of the data cumulation on senders. This issue is tackled by adopting the traffic predictor. A receiver in RISM could smartly poll senders adapting to the variation of senders' traffic loads. By employing the smart polling scheme, the receiver could timely initiate handshakes to reduce the packet queuing delay while constraining the energy consumption on transmitting control packets.

Simulation results show that the performance of RISM with a smart polling scheme outperforms MMAC-CR, a representative CR-based MAC protocol. Specifically, the throughput of RISM is 6× higher than MMAC-CR, while the hop-by-hop delay and overhead of the control packet are only 0.25× and 0.3× of MMAC-CR. Moreover, RISM could work better in a sensor network than in an ad hoc network. The throughput in the former scenario is nearly 2× that in the latter one while maintaining comparable hop-by-hop delay and overhead of control packet.

RISM is a promising system that enables the cognitive technique to work efficiently and in an environment-friendly way in UAN. However, it faces the congestion problem on the CCC.

To mitigate the congestion problem on CCC, we introduced DCC-MAC. One of the most important features of DCC-MAC is that each node could smartly adjust the bandwidth of its control channel by adding a frequency band from the in-band channel when the traffic is heavy, and returning it when the control channel becomes underutilized. The simulation results demonstrated that DCC-MAC could efficiently eliminate the congestion problem of CCC for a cognitive network. In a high traffic generation rate scenario, DCC-MAC provides a ×1.5 and ×4 higher throughput than RISM and MMAC-CR, respectively, while maintaining a comparable or even lower overhead. In addition, the performance of DCC-MAC is not sensitive to the topology of a network. Therefore, it could work in different underwater applications reliably.

References

1. Y. Luo, L. Pu, M. Zuba, Z. Peng, and J.-H. Cui, "Challenges and opportunities of underwater cognitive acoustic networks," *IEEE Transactions on Emerging Topics in Computing*, vol. 2, no. 2, pp. 198–211, 2014.
2. Y. Luo, L. Pu, Z. Peng, and J.-H. Cui, "Dynamic control channel MAC for underwater cognitive acoustic networks," in *Proceedings of the International Conference on Computer Communications (INFOCOM)*. IEEE, 2016.
3. R. Williams, A. J. Wright, E. Ashe, L. Blight, R. Bruintjes, R. Canessa, C. Clark, S. Cullis-Suzuki, D. Dakin, C. Erbe *et al.*, "Impacts of anthropogenic noise on marine life: Publication patterns, new discoveries, and future directions in research and management," *Ocean & Coastal Management*, vol. 115, pp. 17–24, 2015.
4. D. J. Thomson and D. R. Barclay, "Real-time observations of the impact of covid-19 on underwater noise," *The Journal of the Acoustical Society of America*, vol. 147, no. 5, pp. 3390–3396, 2020.
5. C. M. Duarte, L. Chapuis, S. P. Collin, D. P. Costa, R. P. Devassy, V. M. Eguiluz, C. Erbe, T. A. Gordon, B. S. Halpern, H. R. Harding *et al.*, "The soundscape of the anthropocene ocean," *Science*, vol. 371, no. 6529, 2021.
6. N. Baldo, P. Casari, and M. Zorzi, "Cognitive spectrum access for underwater acoustic communications," in *Proceedings of the International Conference on Communications (ICC)*. IEEE, 2008, pp. 518–523.
7. W. Yonggang, T. Jiansheng, P. Yue, and H. Li, "Underwater communication goes cognitive," in *Proceedings of OCEANS*. Quebec City, Canada: IEEE, 2008.
8. W. Richardson and D. Thomson, *Marine mammals and noise*. Academic Press, 1998.
9. M. Wang, A. Agrawal, A. Khandekar, and S. Aedudodla, "Preamble design, system acquisition, and determination in modern OFDMA cellular

communications: An overview," *IEEE Communications Magazine*, vol. 49, no. 7, pp. 164–175, 2011.

10. C. Stevenson, G. Chouinard, Z. Lei, W. Hu, S. Shellhammer, and W. Caldwell, "IEEE 802.22: The first cognitive radio wireless regional area network standard," *IEEE Communications Magazine*, vol. 47, no. 1, pp. 130–138, 2009.

11. Z. Peng, H. Mo, J. Liu, Z. Wang, H. Zhou, X. Xu, S. Le, Y. Zhu, J.-H. Cui, Z. Shi, and S. Zhou, "Nams: A networked acoustic modem system for underwater applications," in *Proceedings of OCEANS*, Hawaii, USA, pp. 1 5.

12. K. Kim, I. Akbar, K. Bae, J.-s. Urn, C. Spooner, and J. Reed, "Cyclostationary approaches to signal detection and classification in cognitive radio," in *Proceedings of the International Symposium on New Frontiers in Dynamic Spectrum Access Networks (DySPAN)*, 2007, pp. 212–215.

13. B. M. Sadler and A. V. Dandawate, "Nonparametric estimation of the cyclic cross spectrum," *IEEE Transactions on Information Theory*, vol. 44, no. 1, pp. 351–358, 1998.

14. E. T. Ar and I. E. Telatar, "Capacity of Multi-antenna Gaussian Channels," *European Transactions on Telecommunications*, vol. 10, pp. 585–595, 1999.

15. C. Y. Wong, R. S. Cheng, K. B. Lataief, and R. D. Murch, "Multiuser OFDM with adaptive subcarrier, bit, and power allocation," *IEEE Journal on Selected Areas in Communications*, vol. 17, no. 10, pp. 1747–1758, 1999.

16. F. F. Digham, "Joint power and channel allocation for cognitive radios," in *Proceedings of Wireless Communications and Networking Conference (WCNC)*. IEEE, 2008, pp. 882–887.

17. S. S. Haykin, *Adaptive filter theory*. Pearson Education India, 2008.

18. B. Yegnanarayana, *Artificial neural networks*. Prentice Hall of India Private Limited, 2009.

19. E. A. Wan, "Finite impulse response neural networks with applications in time series prediction," Ph.D. dissertation, Department of Electrical Engineering, Stanford University, California, USA, 1993.

20. P. Xie, Z. Zhou, Z. Peng, H. Yan, T. Hu, J.-H. Cui, Z. Shi, Y. Fei, and S. Zhou, "Aqua-Sim: An NS-2 based simulator for underwater sensor networks," in *Proceedings of OCEANS*, Biloxi, Mexico, 2009, pp. 1–7.

21. M. Timmers, S. Pollin, A. Dejonghe, L. Van der Perre, and F. Catthoor, "A distributed multichannel MAC protocol for multihop cognitive radio networks," *IEEE Transactions on Vehicular Technology*, vol. 59, no. 1, pp. 446–459, 2010.

22. J. So and N. H. Vaidya, "Multi-channel MAC for ad hoc networks: Handling multi-channel hidden terminals using a single transceiver," in *Proceedings of the International Symposium On Mobile Ad Hoc Networking and Computing (MobiHoc)*. ACM, 2004, pp. 222–233.

23. Y. Luo, L. Pu, Z. Peng, Y. Zhu, and J.-H. Cui, "RISM: An efficient spectrum management system for underwater cognitive acoustic networks," in *Proceedings of the International Conference on Sensing, Communication, and Networking (SECON)*. IEEE, 2014, pp. 414–422.

24. M. Stojanovic and J. Preisig, "Underwater acoustic communication channels: Propagation models and statistical characterization," *IEEE Communications Magazine*, vol. 47, no. 1, pp. 84–89, 2009.

25. Aquasent Company, "AM-OFDM-S1 acoustic modem," aquasent.com. [Online]. Available: http://www.aquasent.com/acoustic-modems/, [Accessed: December, 2014].

26. Teledyne Benthos Incorporation, "ATM-903 acoustic modem," teledynebenthos.com [Online]. Available: http://teledynebenthos.com/product/acoustic_modems/903-series-atm-903, [Accessed: December, 2015].

27. X. Lurton, *An introduction to underwater acoustics: Principles and applications.* Springer Science & Business Media, 2002.

Index

Printed in the United States
by Baker & Taylor Publisher Services